Information Ergonomics

Michael Stein • Peter Sandl
Editors

Information Ergonomics

A Theoretical Approach and Practical Experience in Transportation

Springer

Editors
Michael Stein
Human Factors and Ergonomics
German Air Force Institute of
Aviation Medicine
Manching
Germany

Peter Sandl
EADS Deutschland GmbH
Cassidian
Manching
Germany

ISBN 978-3-642-25840-4 e-ISBN 978-3-642-25841-1
DOI 10.1007/978-3-642-25841-1
Springer Heidelberg Dordrecht London New York

Library of Congress Control Number: 2012934743

Printed on acid-free paper

Springer is part of Springer Science+Business Media (www.springer.com)

Preface

Even in the early stages of mankind individuals and their tribes needed usable information to protect them against threatening environmental factors on the one hand and to be able to seek certain places providing food and survivable climatic conditions on the other hand. Prehistoric man's habitats represented his information system or in other words: information was inherently connected to objects and incidents. It did not appear explicitly. Using his sensory organs, instincts and mental maps he oriented himself in these habitats based on environmental information, e.g. position of the sun/moon, vegetation and soil conditions. His senses and mental models were adapted ideally to his environment for the purpose of survival: to a large extent, his mental models and natural surroundings were congruent. Environmental changes usually occurred gradually so sensory organs and also cognitive and physical resources/capacities could develop and adapt accordingly. Many actions, for instance information search, e.g. for a certain scent indicating rich food or changes in the sky, e.g. as an indication of an approaching thunderstorm were identified on an instinctive basis as a reaction to key stimuli and resulted in corresponding actions, here in search for food. From today's perspective we can rule out any lack of information in this former time. For now, mankind has survived! Perhaps, if man had obtained "better" information, resulting in his supremacy, he might have caused greater damage to himself, nature and other species.

As early drawings show, it seems that it was already at that time when humans wanted to express themselves. In the sense of information systems the character of materials and objects changed. They now carry information, independent from the above-mentioned inherent information content. Rock walls containing images of the daily life, scratched codes in clay tablets or written text on papyrus can be seen as early memories. The information became explicit and emerged visible. For a long time such recorded information could be directly connected with an object carrying the information. Rocks, paper, or in later times, celluloid can be seen as examples.

With respect to this fact the introduction of computers and the respective memory media marked a watershed. The information carriers such as paper tapes or even floppy discs, initially tangible, became more and more a latent part of a machine, e.g. disc drive or even cloud memory in the internet, and are invisible to the user. In that sense the materialization of information is completely abolished. At best, the keyboard and monitor can only be seen as a window into stored information. The original relation of humans to information thus changed completely.

As a consequence, in contrast to prehistoric man, contemporary man has one foot in a virtual reality (VR), which is proceeding to spill over into more and more aspects of life. He is always online; moreover, he is part of the global network and thus has access to all kinds of information systems, whereby the most comprehensive system being the world wide web. As opposed to the natural environment this *artificial* environment is developing at a fast pace, demanding enormous media competence of individuals as well as the intention to participate in these rapid changes. In contrast to a natural environment the mental model and VR do not always match. Since we are not adapted to VR, the reverse conclusion is that VR must be adapted to us (cf. user-centered design). In this context, adaption does not only mean adaption to our perceptive and cognitive abilities and limitations but also to our emotions and motivation, which constitute a highly relevant scope of our experiences and actions.

The factors described above increasingly apply also to information systems in transportation. There are developments, e.g. in aviation, to *complement* the pilot's external view using so-called enhanced vision and synthetic information, thus rendering him less dependent on meteorological conditions. Furthermore, aviation checklists, for instance, are no longer provided in paper versions but rather in so-called electronic flight bags. Within the scope of this development the greatest problem is that the user or operator is inundated with all sorts of irrelevant data (not information!). In this context, irrelevant means that the data has no significance with regard to current or near future problem solutions. Usually, the user or operator does not lack data access. The challenge for him is to find relevant information to solve a certain problem and to identify this information among the diverse accessible data. In addition, information must be comprehensible and also applicable to the problem. However, the initial questions are frequently how the operator can find information: i.e. navigation or means of obtaining information.

The aspects listed are highly relevant, in particular for transportation since the respective systems are time and safety critical. This means that the period of time available for information search and subsequent problem solutions is limited and the potential damage resulting from incidents or accidents is generally very large.

We should also take into consideration that the user is subjected to high stress in this environment which can impede the process of information search and processing. Just think of combat missions of fighter jet pilots. This also explains why so-called declutter modes are installed in modern combat aircraft which enable the pilot to regulate the amount of information displayed. This is the only way he can ensure primary flight control as well as mission conduct in stressful situations. All this shows the importance of designing information systems for the respective purposes and adapting them to user or operator needs.

This book deals with the ergonomic design of information systems in transportation. It is subdivided into a theoretical and a practical part showing examples, attempting to combine both parts in terms of coherence. After all, theory only makes sense if it contributes to sound practical solutions. However, the articles of the individual authors contemplating on the subject of information systems from their respective scientific or practical view have not been "tailored to fit". Information ergonomics is a multidisciplinary field, relying on precision, but also the impact of various technical languages and disciplines.

Since it is the focus of this book, the term information system is defined in the first part. To this end, theories and in particular taxonomies in the field of management information systems are presented and applied to information ergonomics. Furthermore, the significance of information quality is indicated and criteria are cited for the evaluation of information. Subsequently, theoretical insight into information ergonomics are described. In this context the objective and significance are explained as well as the overlap and differentiation of similar fields, e.g. software ergonomics or system ergonomics, in terms of a theoretical deduction of information ergonomics. The chapter on the human aspect of information ergonomics complements the theoretical part of this book. There is a model developed which describes the various phases of mental information processing starting with information access, the decision-making processes during information search as well as the implementation of the information received in corresponding actions. At the same time, this model includes cognitive and motivational as well as emotional processes and should serve as a basis for the design of information systems in transportation. In particular, this should enable an in-depth view of the user side.

The second part counterbalances the theoretical approach by best practices from various sectors of transportation. One chapter highlights information and communication in the context of an automobile driving task. The chapters about civil and military aviation show state-of-the-art developments. The first part stresses the status of available information networks in civil aviation and shows consequences for the workplace of a civil pilot. The latter describes the situation from a military perspective using combat aircraft and so-called unmanned aerial vehicles as examples. Although both disquisitions deal with aviation, it becomes obvious that from an informational perspective the overlap is just marginally. Another chapter devotes specifically to a ground component of aviation, namely air traffic control. Current and future task and work place of an air traffic controller are the central part of that part. Another chapter describes consequences on man–machine-systems in railroading arising from European law.

Prospects on the future development of information systems specifically in the area of traffic and transportation are given at the end of the book.

Manching, Germany

Michael Stein
Peter Sandl

Acknowledgement

We, the editors, would like to express our gratitude to the German Air Force Institute of Aviation Medicine and Cassidian for their continuous support of this book project.

Further, we would like to thank S. Bayrhof, E.-M. Walter and F. Weinand from the Bundessprachenamt for thoroughly translating manuscripts in pleasing chapters as well as Maxi Robinski for attentive proof reading.

At last we wish to deeply express our appreciation to our authors for their valuable contributions. Their articles forms the book as it is.

Contents

Information Systems in Transportation

Alf Zimmer and Michael Stein

1 Introduction

The increasing significance of information technology as well as application networks for varying areas, e.g. aviation and shipping as well as automobile and rail traffic are characteristic traits of an information society (cf. Harms and Luckhardt 2005). Further developments in the fields of microelectronics, computer, software and network technology, but also display technology, etc. are the driving force of this progression.

While the operator was provided with control relevant information via conventional instruments on the one hand and in the form of hardcopies on the other hand, i.e. in a more linear format (e.g. standard flight manual) prior to informatization, current technology enables the storage and presentation of this information[1] in a

[1] Information is "(. . .) that subset of knowledge which is required by a certain person or group in a concrete situation to solve problems and which frequently is not available. (. . .) If it was to be expressed in a formula information is 'knowledge in action'" (Kuhlen 1990, pp. 13 and 14). "This is in line with the psychological concept of information in the framework of "ecological perception" (Gibson 1979); however, Gibson makes the further important distinction between the two modes "direct information pick-up" and "indirect information pick-up", in the first mode data provide the information for action immediately, in the second case data have to be semantically processed in order to guide action. A typical example for direct information pick-up is landing a plane without instruments (Gibson, Olum & Rosenblatt 1955). In the following, the mode of indirect information pick-up will be predominant. However, e.g. the 'intelligent' preprocessing of the data and their display in 3-D can provide a kind of illusory direct information pick-up." In the following the term used in relation to the term "information system" will always be "information", knowing very well that initially "data" are stored in information systems and these data can only become "information" in interaction with the recipient.

A. Zimmer (✉)
University of Regensburg, Regensburg, Germany
e-mail: alf.zimmer@psychologie.uni-regensburg.de

M. Stein
German Air Force Institute of Aviation Medicine, Manching, Germany

M. Stein and P. Sandl (eds.), *Information Ergonomics*,

DOI 10.1007/978-3-642-25841-1_1,

multi-media and hypermedia format. Therefore, the term "multimedia" is defined as follows: "a combination of hardware, software, and storage technologies incorporated to provide a multisensory information environment" (Galbreath 1992, p. 16). If texts, graphics, images, sounds, and videos are provided with cross-reference links, this application becomes a hypermedia application (cf. Luckhardt 2000, p. 1; cf. Issing and Klimsa 1997; Schulmeister 2002; Holzinger 2001; Gerdes 1997).

In addition to a "novel" presentation of information further progress can be observed in the field of transportation: the increasing extent of automation. The human machine interface is currently characterized by both, the changes in the presentation of information (also information access!) as well as a higher degree of automation. The changing processes in aviation can be quoted as examples, which are characterized by the introduction of integrated information displays (multi function displays, glass cockpit concept) as well as enhanced and synthetic vision systems (tunnel in the sky). The progress described can generally be observed with a certain time delay within the automobile sector and in other transport areas.

In addition to informatization the increase of available information results in the mediatization of the described areas. This means that control and action relevant information on system and environment conditions is made accessible to an increasing extent via sensor systems and subsequently presented in media (displays). One example in this regard is the enhanced and synthetic vision system. Via sensor data fusion this system generates an image of the environment surrounding the aircraft, which is augmented by relevant information (enhanced vision). This type of information presentation eventually eliminates the need for pilot external view. Such systems are significant in poor weather conditions in which the real external view yields no reliable information. The aim is an increased situational awareness of the operator, thus ensuring safe operation, while at the same time improving economy. Latter is crucial for transport companies, particularly in times of global trade and also global competition.

However, with regard to the described developments we have to note that the cognitive capacity of man represents the bottleneck with respect to information handling. At the same time, the cognitive resources employed with regard to immensely increasing technical possibilities and also information options for the availability and accessibility of information are nearly the same as prior to the beginning of the "media age" – in spite of improved "education and training". In this context it should be noted that in addition to content knowledge new media also require an increasingly complex strategic knowledge with respect to information retrieval. Furthermore, new media make high demands on communication skills, particularly the capability of differentiating between relevant and non-relevant information. We should also note that new media and the users' media competence are undergoing constant developments and are thus constantly changing. Hence, the user has to dedicate part of his cognitive resources to new learning processes.

In the context of this development the negative effects, e.g. the keyhole effect, tunnelling, fixation, and absorption, should be analysed very precisely in addition to the positive effects like the improvement of the human machine interface by means of a more effective presentation of information (and thus an increase in the safety and reliability of overall systems).

This brief outline encompasses the use of electronic flight bags as detailed in this paper which now provide checklists formerly on paper and other flight and organization relevant information to the pilot as hypermedia-based electronic documentation (information systems). The objective is to provide the pilot with more up-to-date and situation-specific information, improve user friendliness and information accessibility as well as to introduce a paperless cockpit for economic reasons (Anstey 2004).

Referring to these trends Stix (1991) describes the change from traditional cockpits with traditional dial-type instruments to the glass cockpit in his paper titled "Computer pilots. Trends in traffic technology". The role perception of the pilot as an "aviator" is examined critically in this paper. Within the scope of automation the pilot has turned into a system operator, monitoring and controlling the aircraft via multi-function displays and information systems and taking into account economical considerations regarding ideal velocity and altitude in relation to the current personnel costs by using information systems (Haak 2002). The presentation options in the cockpit and the partially resulting data flood in this context frequently cause excessive strain leading to user questions such as: "What's the system doing?" and "What mode is the system in?" (Haak 2002). This demonstrates the significance of an adequate ergonomic design of the presentation of information to improve safety and efficiency. In sum, developments will enable more sophisticated and complex information systems in aircraft, presenting much more than information on primary flight displays (PFD), navigation displays (ND) and system parameters (primary engine display, PED, etc.), as has already been implemented in part. The future belongs to networked, hypermedia-type information systems operated by pilots (system managers) in those work phases in which they are not linked to primary flight controls.

Ramu (2004a, p. 2) as one of the leading researchers of the European Institute of Cognitive Science and Engineering (EURISCO International) states the following in this context: "Once web technologies will enable large industry applications, the knowledge of the whole aeronautical community will be available in a dynamic and interactive way."

2 Definition: Information System

Due to the fact that information systems are applied in various fields (e.g. business data processing, management, library and information science, driver information systems, flight management systems) there is also a large number of specific, partly very pragmatic definitions as it is also the case in the field of transportation. It is, however, difficult to find general definitions and descriptions of the term "information system". Subsequently, we shall attempt to achieve a general definition and description of information systems and describe a driver information system on the one hand and electronic flight bag on the other hand as examples for information systems in the field of transportation.

Business data processing provides a general definition which is also adequate for the field of transportation: "An information system can be defined technically as a set of interrelated components that collect (or retrieve), process, store, and distribute information to support decision making, coordination and control in an organization" (Laudon and Laudon 2008, p. 6).

Kroenke (2008) states that an electronic information system consists of hardware, software, data, procedures, and people. From this perspective information systems are socio-technical systems with the aim of optimum provision of information and communication. These systems are divided into human and machine subsystems. On this basis, Krcmar (2003, p. 47) extends the definition as follows: "Information systems denote systems of coordinated elements of a personal, organizational and technical nature serving to cover information needs."

With regard to hardware, input units (e.g. keyboard, mouse, touch screen) and output units (e.g. computer screen, multi-function displays) can be distinguished. With respect to information systems, Teuber (1999) makes a distinction between application systems (technical components) and the user. The application system consists of hardware and software, with software being divided into basic software and application software.

Apart from the described definitions with the focus of reference on socio-technical aspects Kunz and Rittel (1972) tend to emphasize contextual aspects. With regard to the characterization of the term information system (under the aspect of information system characteristics) the authors focus on the capability of a user to solve a problem within a specific problem context using information systems. "Information systems are provisions which should enable and support external "information" of a user (or a category of users) with regard to a category of his/their problems" (Kunz and Rittel 1972, p. 42).

The following major tasks can be derived from the described definition with respect to designing an information system: a "user analysis" (or classification of users or their characteristics and preferences) as well as an "analysis of problem categories/situations" and their relation (cf. Kubicek et al. 1997).

On this basis Kuhlen (1990, pp. 13–14) states: "Information systems shall provide information to persons in specific problem situations. In this way information systems expand the quantity of knowledge available to human minds. Information is not gained from within oneself or by introspection, information is not remembered. Information is sought from external sources".

An exact description of the problem context, the awareness of the original and target state as well as the operators required for transformation, are of relevance for problem solutions. This is crucial particularly for examining the information gained with regard to its relevance; it is possible that information has to be transferred to the problem situation and stored cognitively until a problem has been solved.

Furthermore, Kunz and Rittel (1972, p. 43) limit the aforementioned definition by declaring the general process of generating information and the provision of information as non-relevant, while qualifying the relation between the generation/provision of information and the support for problem solutions as highly relevant. "A system qualifies as an information system because it should contribute to information and not because it generates or contains information. Only data, which can result in information for the user in a given problematic situation are stored or generated. An information system only includes data which will contribute to external information, i.e. which is not realized as internally generated (e.g. ideas, rethinking processes, etc.). This does not preclude the fact that information systems may not cause internal processes of knowledge diversification (by means of external stimuli)".

In summary, indicating the limits of information systems, Kunz and Rittel (1972, p. 42) state: "the goal of the design of information systems ultimately is to make provisions for a category of problematic situations an actor or a category of actors expect to encounter. Information systems are therefore always set up for categories of problematic situations and frequently for categories of actors." Contrary to this ideal perspective every problematic situation can be seen as unique. To this end, no information system has the capability to provide the total information to solve a problem.

3 Criteria for the Assessment of Information Systems: Data and Information Quality

Data and information quality are crucial to the quality of an information system, regardless of its application range. The terms data quality and information quality are frequently used as synonyms in publications in the field of business data processing. In this context data quality covers technical quality (e.g. hardware, network), while information quality (information access) covers non-technical aspects (Madnick et al. 2009,[2] see also Helfert et al. 2009). Furthermore, Wang et al. (1995) criticize the fact that there are no consistent definitions of the terms "data quality" and "information quality" in scientific research. Price and Shanks (2005) as well as Mettler et al. (2008) put this down to the fact that there are four different scientific approaches with regard to the definition of data quality and information quality: "empirical, practical, theoretical and literature based" (Mettler et al. 2008, p. 1884).

In an article giving an overview of database (information) quality Hoxmeier (1997a, b) states that numerous approaches have been developed over time in the attempt to set up "factors, attributes, rules or guidelines" for the evaluation of information quality (Martin 1976; Zmud 1978). In this context the authors established a number of evaluation dimensions. Various studies (Helfert et al. 2009, p. 1) also show that information quality is a multidimensional concept. So far, the scientific community has not agreed on a consistent definition.

---

[2] With regard to the publication by Madnick et al. (2009) no distinction will be made between technical and non-technical aspects within the scope of this publication.

One of the early approaches of evaluating computer based information systems and information quality was established by Martin (1976). He proposes the following 12 evaluation criteria:

- "Accurate,
- Tailored to the needs of the user,
- Relevant,
- Timely,
- Immediately understandable,
- Recognizable,
- Attractively presented,
- Brief,
- Up-to-date,
- Trustworthy,
- Complete,
- Easily accessible".

With regard to information quality (IQ) Wang and Strong (1996, cf. Strong et al. 1997) differentiate between the following:

- "Intrinsic information quality: Accuracy, Objectivity, Believability, Reputation
- Contextual information quality: Relevancy, Value-Added, Timeliness, Completeness, Amount of information
- Representational information quality: Interpretability, Ease of understanding, Concise representation, Consistent representation
- Accessibility information quality: Accessibility, Access security".

Based on the research by Wang and Strong (1996) the "Deutsche Gesellschaft für Informations- und Datenqualität" (German Society for Information and Data Quality[3]) has derived the following criteria:

- "Accessibility (information is accessible if it is directly available to the user by applying simple methods),
- Appropriate amount of data (the amount of data is appropriate if the quantity of available information fulfills the requirements),
- Believability (information is considered believable if certificates prove a high standard of quality or considerable effort is directed to the acquisition and dissemination of information),
- Completeness (information is complete if not missing and available at the defined times during the respective process steps),
- Concise representation (information is concise if the required information is presented in an adequate and easily understandable format),

[3] Committee: "Normen und Standards der Deutschen Gesellschaft für Informationsqualität" (Standards of the German Society for Information Quality, http://www.dgiq.de/_data/pdf/IQ-Definition/IQDefinitionen.pdf. Accessed 28th January 2010).

- Consistent representation (information is presented consistently if displayed continuously in the same manner),
- Ease of manipulation (information is easy to process if it can be modified easily and used for various purposes),
- Free of error (if information is conform with reality),
- Interpretability (interpretation of information is unambiguous and clear if understood to have identical, technically correct meaning),
- Objectivity (information is considered objective if it is purely factual and neutral),
- Relevancy (information is relevant if it provides necessary information to the user),
- Reputation (information is valued highly if the source of information, transport media and the processing system have a high reputation of reliability and competence),
- Timeliness (information is considered current and up to date if the actual characteristics of the described object are displayed in near real time),
- Understandability (information is understandable if directly understood by the users and employable for their purposes),
- Value-added (information is considered to be value-added information if its use can result in a quantifiable increase of a monetary target function)".

Since this list refers to business data processing we have to assess very carefully which criteria are relevant to the field of transportation.

In the context of transportation the objective of information systems (information) is the support of the operator in specific problem categories within the scope of system control to contribute to finding a solution. Since a successful solution depends on

- The operator (knowledge, training – novice, experienced, expert, etc.),
- The information system (receivability of information, design of information, adaptivity),
- The system status (e.g. complexity of the failure) as well as
- The environmental conditions (complexity),

the neutral information content can be determined only empirically for certain operator populations as well as problem and situation categories (cf. Information Science). In this context quality and significance of information depend on their access and contribution to the problem solution. This can be simulated for instance under experimental conditions in the flight simulator. Abnormal situations, for example, can be simulated in certain flight phases during which the support of adaptive warning panels or electronic flight bags is used as a problem solution.

When considering the phases of information access (see Stein 2008) and the implementation of information under the aspect of system control (or also system monitoring) the subjectively perceived quality of information depends on the objectives, expectations, capabilities (operation of the information system), specialist knowledge and the field of application (e.g. system knowledge with regard to the aircraft, scope and quality of the problem, etc.) and the perceived characteristics of information and information system (user-friendly design, concise, conform to expectations). There are the following expectations of information characteristics:

- Duration of information search,
- Scope of information,
- Context sensitivity of information (information system adapted to the dynamic system and environmental conditions),
- Time of availability of information,
- Presentation of information,
- Cost benefit of information,
- Interreferential information.

The terms "recall" and "precision" can be used to determine the quality of information systems. The first index includes the ratio between the relevant documents found within an information system and the total number of documents. The second index describes the number of documents found which are at all relevant. The values for recall and precision range between 0 and 1. The closer the index is to 1 the more relevant documents were found Luckhard (2005).

4 Context of Use Regarding Information Systems

4.1 Electronic Flight Bag

More and more airlines replace traditional paper-based documentation (e.g. emergency checklists, Instrumental Flight Rules, aircraft and flight operations manuals) by electronic documentation presented and stored in so-called electronic flight bags. Examples in this context are Lufthansa and LTU (Skybook), Austrian Airlines (Project Paperless Wings) as well as GB Airways (Teledyne Electronic Flight Bags). The objective is a paperless cockpit with all related economic savings (less update effort, etc.), possibly improved user friendliness and information accessibility (Anstey 2004). The use of electronic flight bags shall provide the pilot with more up to date and situation adequate information, which should also contribute to enhance flight safety. In this respect Ramu (2004a, p. 1) postulates the following: "The final role of operational documentation is to help the end user to access the right information at the right time, in the right language, and the right media at the right level of detail using the right device and user interface".

Chandra et al. (2004) mention the low costs compared to traditional avionics applications on the one hand and the vast range of functionalities and their flexibility and thus also their application in various different flight sectors, e.g. charter/ business, air transport or regular airline transport on the other as reasons for the rapid development and increased use of electronic flight bags. These have the potential for savings in this sector, which is characterized by intense competition, thus possibly resulting in a competitive edge.

In the field of military aviation checklists are also partly stored electronically and made available to pilots in this way. Based on operational conditions, especially of combat aircraft (G forces, bearing) in military missions Fitzsimmons (2002), however, states that portable equipment, e.g. handhelds as used in civilian aviation, is suitable only to a limited extent for military missions. This type of equipment poses a threat to safety, e.g. during emergency egress or also due to possible sun reflections.

4.1.1 Definition and Classification of Electronic Flight Bags

For the field of civil aviation the term electronic flight bag is defined in the Advisory Circular of the Federal Aviation Administration (FAA AC 120-76A 2003, p. 2) as follows: "An electronic display system intended primarily for cockpit/flight deck or cabin use. EFB devices can display a variety of aviation data or perform basic calculations (e.g., performance data, fuel calculations, etc.)". The following description is also included: "(. . .) In the past, some of these functions were traditionally accomplished using paper references or were based on data provided to the flight crew by an airline's "flight dispatch" function. The scope of the EFB system functionality may also include various other hosted databases and applications. Physical EFB displays may use various technologies, formats, and forms of communication. These devices are sometimes referred to as auxiliary performance computers (APC) or laptop auxiliary performance computers (LAPC)".

The US Air Force Materiel Command which is concerned with the transformation of paper-based technical orders (TOs for the technical maintenance of aircraft) and also the conversion of checklists (Flight Manuals Transformation Program [FMTP]) into electronic documents also defines electronic flight bags as follows: "A hardware device containing data, previously available in paper format (flight manuals, electronic checklists (ECL), Flight Information Publication (FLIP), Specific Information (SPINS), AF Instructions, TPC charts/maps, etc.), required to operate and employ weapon systems. A more realistic role of the EFB is supporting information management. Information management attempts to support flexible information access and presentation so that users are enabled to more easily access the specific information they need at any point in the flight and to support effective, efficient decision making thus enhancing situational awareness" (Headquarters Air Force Material Command, Flight Manuals Transformation Program Concept of Operations 2002, p. 15).

In particular, this extended definition highlights the options of managing and employing force systems more adequately using electronic flight bags. The definition also refers to information management, which provides the pilot with up to date information during each flight phase. Furthermore, the use of electronic flight bags should make decision-making processes more efficient and effective, thus supporting situational awareness of the pilot.

The Advisory Circular of the Federal Aviation Administration (2003) differentiates between three hardware and three software categories of electronic flight bags:

- Hardware category 1 includes commercial off-the-shelf, portable personal computers, which are not permanently installed in or connected to other aircraft system units. There are no mandatory certification processes for this type of equipment. Products of this category shall not be used in critical situations.
- Hardware category 2 is also characterized by commercial off-the-shelf transportable PCs, however, these are connected to specific functional units in the aircraft during normal operations. Certification processes are mandatory for this category.
- The third hardware category comprises permanently installed electronic flight bags (multi-functional displays). These are subject to certification processes.
- Software category A consists of flight operations manuals and company standard operating procedures. An approval by the Flight Standard District Office is necessary for this category.
- Software category B includes more effective programs, which can be used for calculations, etc. (weight and balance calculations, runway limiting performance calculations). An approval by the Flight Standard District Office is necessary for these programs.
- Software category C comprises programs subject to certification, which are used for determining and displaying the one's own position in maps which are coupled to other systems (own ship position). However, these must not be used for navigation.

Due to the fact that electronic flight bags represent a rather new accomplishment their practical use in the coming years will decide which of the categories described will best cover the requirements of various aircraft applications (general and military aviation) and which will be successful on the market.

4.1.2 Composition and Structuring of Electronic Flight Bags

In the course of the development of electronic flight bags the available paper based documentation was initially converted on a "one-to-one basis" into electronic documentation (Barnard et al. 2004). In this respect Ramu (2002) states that typical characteristics of paper-based documentation, e.g. outlines in chapters or linear coherence generated by subsequent pages, will be lost in electronic documentation. New forms of presentation ("characteristics") should be defined.

Questions in this context are for example the following:

- How to design information hierarchy and access (cf. system approach or task approach, Ramu 2002),
- The coexistence between paper based and electronic documentation,

- Prioritization to support the simultaneous management of various databases/ documents,
- Navigation to required information,
- And the security of e-documents.

Hence, one of the most important questions with regard to the development of electronic flight bags is the "ideal method" of transformation of traditional paper-based documentation – for flight operations as well as maintenance – to electronic documentation. Anstey (2004), an expert in electronic documentation applications at Boeing AG, proposes an initial analysis of available paper based documentation with respect to its composition and structuring (chapters, structured vs. non-structured), format (Word, pdf-file, etc.) and the contents presented. The following are subsequent questions:

- "Does it make sense to adopt the given structure of the paper-based version in electronic documentation?"
- "Is the organization of the documentation adequate for display presentation?"
- "How many information levels should the electronic documentation have?"
- "Can tables, diagrams and graphics be integrated adequately – and in particular readably – in the electronic documentation?"

Furthermore, Anstey (2004) postulates that it is essential to take into account the know-how of the end users (pilot) and their experience and capabilities for an adequate design of electronic flight bags. The following are relevant from an ergonomic point of view:

- Experience and knowledge of the various flight decks,
- Frequency of use of various manuals,
- Identification of flight phases during which specific manuals are used,
- Organization, structure and composition of electronic flight bags,
- User access to contents,
- Navigation instruments within electronic flight bags for information retrieval,
- Links,
- Integration in flight deck data,
- Presentation of tables and graphics,
- Identification of non-relevant information.

Barnard et al. (2004) postulate that the presentation of documentation (paper-based vs. electronic) will not only change, but it rather will become less important for flight operations. They list the following reasons:

- Required data (information) is integrated in existing systems (displays),
- More independent (self-evident) systems,
- Improved links to ground stations and thus also improved support,
- Increasing complexity and automation of systems (resulting in the fact that the capabilities of the users as well as the requirement to understand the systems by reading documentation during flight will decrease),

- "(...) ubiquitous computing, making available any information anywhere and just-in-time, not necessarily contained in a specific device in a specific form" (Barnard et al. 2004, p. 14).

Some of the arguments, e.g. the "integration of additional information in existing subsystems" (multi-function displays) seem plausible, however, fundamental considerations concerning the design of information systems (structure and organization) in accordance with human factors criteria will still be necessary. Other arguments, e.g. the "ubiquitous availability of information" have already proven false in other highly computerized sectors since it is not the availability of information which is crucial to its usability, but rather the extent of information which can be processed cognitively by the user.

The use of documentations will also be undergoing changes. In this context, Barnard et al. (2004) differentiate between use during flight and use during training. During the flight phase information is required for the selection of tasks (decision making), anticipation of tasks and their effects, sequencing of tasks as well as performance of tasks. Further fields of use include preparation and debriefing phases comprising e.g. simulations or representing specific weather conditions or also flight scenarios/manoeuvres. In the field of training electronic documentation is used as a source of declarative knowledge as well as for computer-based training or also as a dynamic scenario generator.

Ramu (2002) established one method for the user-oriented structuring of electronic documentations, which was also discussed and implemented within the scope of the A 380 development. As demonstrated by the example of Flight Crew Operating Manuals (FCOM) the basic idea is to relate the structuring of electronic documentations to the sequence of tasks to be performed during flight. Task-based documentation takes into account the fact that the work (tasks and subtasks) of a flight crew during flight is time-critical and limited within a specific timeline. Since the cognitive structure of the end user (pilot) is conform also to this timeline and the sequential task requirements it makes also sense for the structure of electronic documentation to be oriented to this critical timeline. This means that the respective relevant information is attributed to the successive tasks for the purpose of "user friendly" information retrieval.

The structuring process is based on the concept of the Documentary Units (DU), which includes the segmentation of all available data in so-called entities. A documentary unit contains data, e.g. descriptions, schematics, animations or also performance data, etc. on the one hand and documentary objects associated with style sheet metadata properties, e.g. colour, type size, spacing, etc. on the other (cf. Ramu 2004a, b). Documentary units are hierarchically organized in sub-documentary units. These are classified in specific user-oriented domains. For the Flight Crew Operating Manual (FCOM) these are subdivided into the following user-oriented domains: Procedures, descriptions, limitations, performances, supplementary techniques, and dispatch requirements. For the optimization of the information retrieval process so-called descriptors – metadata – are assigned to each documentary unit; these are domain descriptors, context descriptors, task descriptors, etc. Their task is to describe the respective contents of the (self-sufficient) documentary units and their contexts.

Ramu differentiates between three categories of descriptors in aviation:

- The artefact family (system [e.g. hydraulics] and interface [e.g. overhead panel]),
- The task family (phases of flight [takeoff, cruise, landing] and operation [e.g. precision approach]),
- The environment family (external and internal environment).

The task-oriented structuring of electronic documentations is based on the task block concept. Ramu (2002) describes the concept as follows: "A task can be represented as a Task-Block, taking into account its context history through Task Descriptors (TD), and leading to a set of actions (Ramu 2002, p. 2)".

Ramu proceeds to introduce the concept of task block organization consisting of several data layers. He uses the image of a house as an analogy. The house symbolizes the highest layer, e.g. flying a specific aircraft type from one location to another (house task descriptor). The second layer of task descriptors includes the successive flight phases of preparation, departure, cruise and arrival. The individual flight phases can be considered to be rooms. There are so-called Standard Operation Procedures (SOP) for each one of these rooms.

In his later work (Ramu 2004a, b, p. 3) he introduces the meta-concept of ontology for classification (see Fig. 1), referring to a definition by van Heijst (1995). "An ontology is an explicit knowledge-level specification of a conceptualisation, e.g. the set of distinctions that are meaningful to an agent. The conceptualisation – and therefore the ontology – may be affected by the particular domain and the particular task it is intended for".

With regard to the structure and composition of ontologies Ramu (2004a, b) further explains: "An ontology dimension will take the form of a logical tree, with one root, multiple nodes hierarchically organised through multiple branches. Root and nodes are context descriptors".

The aforementioned task family and environment family are linked in various ways. This fact is illustrated by the example of poor visibility (environment family) affecting the landing and approach flight phases (task family). The operation category is subdivided into n operations to enable the performance of the tasks required in each flight phase. In this context each descriptor of an operation category is linked to an action to be conducted. The first ontology dimension is defined as the standard, representing all standard actions. If, for instance, the ontology dimension environment family changes the task family including its related actions are also affected. The individual ontology dimensions are independent of each other, meaning that the accounts A and B, for instance, are not connected (Fig. 2).

The briefly described approach to information structuring and information retrieval in electronic flight bags (the connection of a documentation to the sequence of tasks to be performed during a flight as well as the context sensitivity generated, etc.) can be described as an important step towards a user-oriented

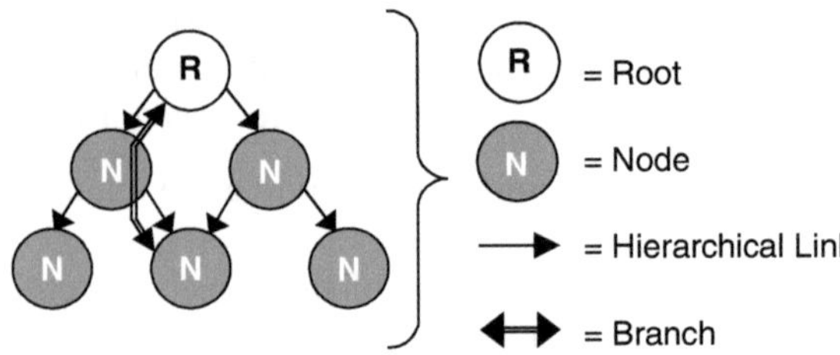

Fig. 1 Ontology dimension properties (Ramu 2004a, b)

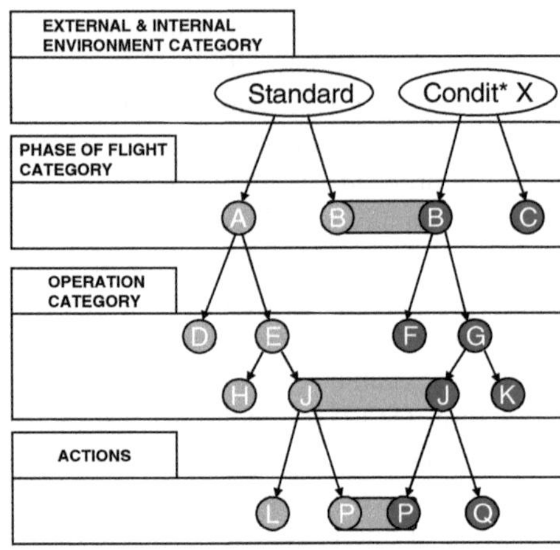

Fig. 2 Task and environment family ontology (Ramu 2004a, b)

design of electronic flight bags. However, future information will not only be presented "rigidly", i.e. automatically, in accordance with specific flight phases using electronic flight bags, but the physical or psychological conditions of pilots (e.g. cognitive state) will also be taken into account with regard to the automatic selection of information. In addition, with the continuous improvement of sensor systems, "external factors" (the environment) will play an increasing role in the automatic presentation of information.

Next to information preselection, information access via searches and browsing is still relevant. We can also assume that there is a trend towards hypermedia structuring of information with regard to the presentation of information and thus similar design guidelines will apply as for other hypermedia information systems.

4.1.3 Checklist for the Assessment of Human Factors

Characteristics of Electronic Flight Bags

With regard to the assessment of electronic flight bags the Advisory Circular of the Federal Aviation Administration (2003) explicitly indicated the requirement to assess electronic flight bags under the human factors aspect. However, no assessment procedure was specified (Chandra et al. 2004). A list of the most relevant human factors topics, e.g. interface design, readability, fault tolerance and workload, etc. has been published as a guide. Based on previous recommendations and studies (Chandra and Mangold 2000; Chandra 2003, etc.) Chandra (2003) developed an evaluation tool for electronic flight bags. The target group of this tool consists of FAA evaluators on the one hand and system designers on the other, who can assess the conformity of the designed systems with regulations using the tool during the development process prior to the actual FAA evaluation. The addressees are explicitly non-human factors experts. Chandra et al. (2004) developed two checklists: one checklist on a meta level including 20 items and one specific checklist with 180 items, with the authors preferring the shorter version for test economical reasons: "We quickly realized that a short paper-based tool that could serve as a "guide for usability assessment" would be more practical" (Chandra et al. 2004). Contents of the short version include symbols/graphic icons (readability, understandability of designation, etc.), formatting/layout (type, font size, etc.), interaction (feedback, intuitive operation), error handling and prevention (error messages, error correction, etc.), multiple applications (recognition of own position within the system, changes between applications), automation (user control over automation, predictability of system reaction), general contents (visual, auditive, tactile characteristics of the system, color application, consistency) workload (problem areas), etc.

Within the scope of tool development several evaluations were performed, e.g. regarding understandability or also coherence of the items. To this end, teams of two evaluators each with different backgrounds – human factor experts, system designers, licensed pilots – assessed real systems and expressed their thoughts by means of the think-aloud method. The reason that the authors cite for this procedure is that a team of two evaluators will usually detect more tool errors than two people processing the system with the tool one after the other. Since the subjects were not part of the future target group they were handed out the basic document of the Advisory Circular of the Federal Aviation Administration (AC 120-76A) to get a feeling for the target group.

The procedure of the evaluation was as follows:

- Introduction (15 min),
- Task-based exploration (90 min),
- Instrument processing (short and long version, 60 min each) and
- Feedback (15 min).

The comments of the evaluators were assessed separately for the task-based exploration and the tool-based review; the data, however, could not always be assessed separately. The comments were then additionally categorized into three priorities and evaluated with regard to frequency. The tool was then modified in accordance with the results of the evaluation. Items with only minor reference to the human factor aspect were sorted out or reformulated if not understandable. Furthermore, in view of the fact that system designers make specific requirements for the evaluation tool, e.g. support of the workflow, design changes were made.

All in all, the tool for the evaluation of electronic flight bags (Chandra et al. 2003) can be used as a screening instrument in a time- and cost-saving manner to indicate “major” human factors errors. As noted by Chandra (2003) the tool cannot replace regular human factors and usability tests. The downsides are that errors can indeed be identified using the tool, however, no solutions or approaches are provided.

4.2 *Driver Information System*

In contrast to air, rail, or sea traffic, road traffic is characterized by a comparable lack of specific rules and very few standard procedures – plus a nearly complete lack of any kind of centralized oversight and control. That is, the amount of possible situations and the resulting demands on decisions and behaviour of drivers exceeds the possibility of regulations for any specific situation.

For this reason, the German law regulating traffic behaviour (StVO) starts with a very general statement concerning the obligations which all traffic participants must meet (§1): In all situation the behaviour has to be such that no other participant is endangered or unduly impeded. Similar general (“catch all”) formulations can be found in the laws regarding traffic behaviour of most countries. The necessity for such a general obligation of due diligence in traffic behaviour can be seen in its importance for most traffic related litigations.

The consequence of this regulatory situation in road traffic is that the responsibility of the individual driver is of utmost importance – and not that of centralized institutions as in air traffic control. For this reason, information systems for road traffic have to satisfy the needs of the driver in specific situations (in the following, only the term driver is used because information systems for other traffic participants still are in early stages of development and distribution). Of utmost importance, however, is that the interaction with these information systems does not only not interfere with the main task “driving”, but that the quality of the main task is improved by the information system (Dahmen-Zimmer et al. 1999). This is the core of the assistive technology approach to road traffic (Wierwille 1993).

In contrast to the guidelines for information systems cited in the first part of this chapter, information systems for road traffic are quite often

- Not integrated, segmental or insular
- Incomplete
- Not up to date and
- Inaccurate or incomplete.

This situation is due to the high complexity of road traffic with many independent agents, extremely variable situations, and extremely short response times. A general framework for such settings has been suggested by Suchman (1987); the concept of ‘situated action’ systematizes the interrelations between plans, situational constraints and opportunities (Dahmen-Zimmer & Zimmer 1997), and available resources.

Thus, the very nature of road traffic prohibits the development of general information systems which follow the cited guidelines for good information systems, instead, it necessitates supportive and assistive information systems which enable the driver to behave responsibly, that is, to minimize danger (and improve safety) and to contribute to a more effective traffic flow.

The driving task requires a whole range of modes of behaviour, which can be classified on three levels: Regulating (e.g. lane keeping), manoevring (e.g. overtaking), and navigating (e.g. goal setting; Verwey & Janssen 1988).

The time available for information processing in traffic is usually very short, typically

- 250–400 ms for regulation (e.g. lane keeping, keeping the necessary distance to other traffic participants, or adjusting the speed according to a traffic sign)
- 1,000–3,000 ms for manoeuvring (e.g. doing a left turn, switching the lane, or overtaking)
- 10,000 ms and above for navigating (e.g. choosing the right roads for attaining a set destination, taking a deviation in case of traffic congestion, or searching for service).

For the support of regulating behaviour, information systems have to be extremely fast, rely on incomplete or fuzzy data, and therefore their output for the driver has to be in a modality compatible with the motor behaviour (optimally tactile) and in a way which is advisory for the driver and not compulsory (see Zimmer 2001). There exist assistive systems for lane keeping, distance warning, and traffic sign recognition – all with more or less detailed information systems in the background. However, the question concerning the best mode of feedback is still open. The general consensus is that the modality ought to be tactile in order to support the motor behaviour optimally but the optimal mode for giving information about speeding or deviating from the lane (e.g. force feedback in the steering wheel or the accelerator, vibrations, etc.) has still to be found, even conventions are missing (Zimmer 1998; Bengler 2001; Bubb 2001; Grunwald and Krause 2001).

The intelligent support for manoeuvring is most challenging because highly dynamic scenes with many independent agents have to be analyzed and evaluated in a very short time (Rumar 1990). In order to develop efficient assistive information systems for this level of driving, reliable car-to-car communication would be necessary, informing not only about the exact position, but also speed, and destination of any car in the environment in order to compute possible collision courses or to advise for an optimal traffic flow. The existing collision warning or prevention systems are therefore confined to scenes of low complexity, that is, with few agents, and low dynamics, that is, they are most efficient with static obstacles.

Only for navigating there exist information systems in the form of digitized maps, which come close to the above mentioned guidelines for information systems. In combination with wireless digital traffic information these systems provide good support for the optimization of way finding. The major short comings of the digital maps underlying the navigation systems are due to the fact that they do not implement recent and actual data regarding road construction or maintenance and the information about traffic signs and regulations is updated only annually. These shortcomings are well known and have been addressed in research programs and guidelines of the EU but as long as the communal, state, and Europe wide data bases for the actual state of the road systems are not compatible a host of reliable and important data cannot be implemented into the navigation systems.

A further drawback for the users of these navigation systems lies in the fact that actual driving data are not collected and used for making the system more up to date. That is, information leading to inconvenient or even false suggestions for routes cannot be corrected because it is not possible for the driver to give feedback about obvious failures to the system. And the system itself does not register the fact that, for instance, a driver repeatedly and systematically deviates from the suggested routes.

In the beginning of this section, the lack of centralized oversight or control in road traffic has been mentioned. However, one exception for this should be noted: in the field of logistics there exist GPS-based systems of fleet control which allow not only just-in-time delivery but also improve the efficiency of the fleet. However, these systems are usually confined to single providers for supply chain management. The competitive edge for any logistics provider is to be better than the others; therefore exchange of information between corporations is scarce. Thus, the existing information does not improve the overall efficiency of the road traffic system.

5 Further Developments

In order to chart the further development of user oriented information systems for road traffic it is necessary give a systematic overview of the user's demands: **what** is needed **when** and **how** to be known by the user, that is, beyond what is obvious or already known by the user. Future systems have to address the following main topics: situational and individual specificity of information, actuality and relevance of the information, and effectivity of the mode of transmitting this information to the user.

Due to the described complexity of events in road traffic, centralized information systems beyond the existing digital maps will not be possible. The focus will be directed on information systems supporting the individual driver in very specific situations. Such systems ought to be local, in the extreme case covering only the traffic in a certain junction. If it is possible for the area of interest to localize all traffic participants, to determine their velocity and direction, and to project the most probable trajectories, an optimal traffic flow can be computed and corresponding information for the guidance of individual drivers can be provided.

However, the effectivity of such a local traffic management system will depend on the willingness of the drivers to comply with the suggestions given by this system. Partially this compliance can be enforced, e.g. by traffic lights and adaptive traffic regulations, but – at least for the foreseeable future – it will not be legally possible to directly influence acceleration and deceleration of the cars in this situation. This has to be done by communicating suggestions to the drivers how to manoeuvre their cars. The quality of such a local system would be enhanced if all cars are equipped with car-to-car and car-to-system communication. For the near future this will not be the case and therefore the necessary information has to be determined by high-speed video observation and analysis. The communication about the effective traffic flow via the traffic management system has to be very fast because any delay of relevant messages will influence the trust into this system and therefore diminish the willingness to comply with the guidance information.

In order to provide not only local support and a local improvement of traffic flow but also an overall improvement, the local information systems have to be designed in such a way that they can interact with other local information systems and higher level traffic management systems in order to prevent traffic congestions on a larger scale. This kind of regional guidance information is already provided by digital traffic information messages, which interact with in-car navigation systems. However, the lack of actuality and reliability of such information blocks compliance in most cases.

From our point of view, the future of information systems for traffic will lie in the development of situation specific but interconnected systems, providing information that fits the actual demands of the driver. In order to achieve this goal, further investigations of time structure and interaction patterns of traffic will be necessary for a better match between what the driver observes in a specific situation and what information is provided for a better and safer traffic flow. As long as enforced compliance is neither technically feasible, nor legally admissible, traffic information systems have to enhance and assist the competence of the driver and not to govern or override it.

The effective mesh of driver competence and information support will depend on the quality of the information given, its specificity for the situational demands, and the fit of the mode of informing and the resulting actions.

The challenge for information science will be the communication between the systems on the different levels, from the systems providing information for the lane keeping of a single car, to the regional or even more general traffic management. If this effective communication between systems can be achieved, it will be possible to reach more safety (see the “Vision Zero”: "the goal for road traffic safety proposed by the commission of the European Union") and effectivity.

References

Akyol, S., Libuda, L., & Kraiss, K.F. (2001). Multimodale Benutzung adaptiver KFZ-Bordsysteme. In T. Jürgensohn & K. P. Timpe (Eds.), *Kraftfahrzeugführung* (pp. 137–154). Berlin: Springer.

Anstey, T. W. (2004). Electronic documentation and the challenge of transitioning from paper. *Proceedings of the HCI-Aero. Workshop electronic documentation: Towards the next generation*, EURISCO International, Toulouse.

Barnard, Y., Ramu, J.-P., & Reiss, M. (2004). Future use of electronic manuals in aviation. *Proceedings of the HCI-Aero. Workshop electronic documentation: Towards the next generation*. EURISCO International, Toulouse.
Bengler, K. (2001). Aspekte multimodaler Bedienung und Anzeige im Automobil. In T. Jürgensohn & K. P. Timpe (Eds.), *Kraftfahrzeugführung* (pp. 195–205). Berlin: Springer.
Blattner, M., Sumikawa, D., & Greenberg, R. (1989). Earcons and icons: Their structure and common design principles. *Human Computer Interaction, 4*, 11–44.
Bubb, H. (2001). Haptik im Kraftfahrzeug. In T. Jürgensohn & K. P. Timpe (Eds.), *Kraftfahrzeugführung* (pp. 155–175). Berlin: Springer.
Chandra, D. C. (2003). A tool for structured evaluation of electronic flight bag usability. *Proceedings of the 22nd digital avionics systems conference*, Indianapolis, pp. 12–16.
Chandra, D. C., & Mangold, S. J. (2000). Human factors considerations in the design and evaluation of electronic flight bags (EFBs) Version 1: Basic functions. Report No DOT-VNTSC-FAA-00-2 MA USDOT. Cambridge: Volpe Center.
Chandra, D. C., Yeh, M., & Riley, V. (2004). Designing a tool to assess the usability of electronic flight bags (EFBs). Report No DOT/FAA/AR-04/38 MA USDOT. Cambridge: Volpe Center.
Chandra, D. C., Yeh, M., Riley, V., & Mangold, S. J. (2003). Human factors considerations in the design and evaluation of electronic flight bags (EFBs). http://www.volpe.dot.gov/hf/aviation/efb/docs/efb_version2.pdf. Accessed 16 Aug 2011.
Dahmen-Zimmer, K., Huber, M., Kaiser, I., Kostka, M., Scheufler, I., Pieculla, W., Praxenthaler, M., & Zimmer, A. (1999). *Definition und Validierung von Kriterien für die Ablenkungswirkung von MMI-Lösungen*. Endbericht MoTiV – MMI AP/5, Regensburg.
Dahmen-Zimmer, K., & Zimmer, A. (1997). *Situationsbezogene Sicherheitskenngrößen im Straßenverkehr*. Bergisch Gladbach: BASt.
Fang, X., & Rau, P. L. (2003). Culture differences in design of portal sites. *Ergonomics, 46*, 242–254.
Federal Aviation Administration Advisory Circular (FAAC). (2003). Guidelines for the certification, airworthiness, and operational approval of electronic flight bag computing devices. Report No FAA AC 120- 76A. http://www.airweb.faa.gov/Regulatory_and_Guidance_Library/rgAdvisoryCircular.nsf/0/B5DE2A1CAC2E1F7B86256CED00786888?OpenDocument&Highlight=efb. Accessed 21 Aug 2011.
Fischhoff, B., MacGregor, D., & Blackshaw, L. (1987). Creating categories for databases. *International Journal of Man–Machine Studies, 27*, 242–254.
Fitzsimmons, F. S. (2002). *The electronic flight bag: A multi-function tool for the modern cockpit*. Institute for Information Technology Applications. United States Air Force Academy Colorado IITA Research Publication 2 Information Series. http://www.usafa.edu/df/iita/Publications/The%20Electronic%20Flight%20Bag.pdf. Accessed 16 Aug 2011.
Galbreath, J. (1992). The educational buzzword of the 1990's: Multimedia, or is it hypermedia, or interactive multimedia, or...? *Educational Technology, 4*, 32.
Geer, D. (2004). Will gesture-recognition technology point the way? *Computer, 37*, 20–23.
Gerdes, H. (1997). Lernen mit Text und Hypertext. Pabst Science Publishers.
Gibson, J. J. (1979) The ecological approach to visual perception. Boston: Houghton-Mifflin.
Gibson, J. J., Olum, P. & Rosenblatt, F. (1955) Parallax and perspective during airplane landings. American Journal of Psychology, *68*, 372–385.
Grunwald, M., & Krause, F. (2001). Haptikdesign im Fahrzeugbau. In M. Grunwald & L. Beyer (Eds.), *Der bewegte Sinn. Grundlagen und Anwendungen zur haptischen Wahrnehmung* (pp. 171–176). Basel: Birkhäuser.
Haak, R. (2002). Das Human Faktor Concept in der Luftfahrt. *Technische Mitteilungen (Arbeitsschutz und Arbeitssicherheit), 95*(2), 104–116.
Harms, I., & Luckhardt, H.-D. (2005). Virtuelles Handbuch Informationswissenschaft. Der Gegenstand der Informationswissenschaft. http://is.uni-sb.de/studium/handbuch/kap7.html. Accessed 18 Aug 2011.

Headquarters Air Force Materiel Command. (2002). *Flight manuals transformation program concept of operations*. Ohio: Operations Division, Directorate of Operations, Wright-Patterson AFB.
Helfert, M., Foley, O., Ge, M., & Cappiello, C. (2009). Analysing the effect of security on information quality dimensions. *The 17th European conference on information systems*, Verona, pp. 2785–2797.
Hoxmeier, J. (1997a). A framework for assessing database quality. Proceedings of the workshop on behavioral models and design transformations: Issues and opportunities in conceptual modeling. *Proceedings of the ACM 16th international conference on conceptual modeling*, Los Angeles.
Hoxmeier, J. A. (1997b). A framework for assessing database quality. *Proceedings of the association of computing machinery ER 97 conference*.
Issing, L. J., & Klimsa, P. (1997). *Information und Lernen mit Multimedia*. Weinheim: Beltz Psychologie Verlagsunion.
Johannsen, G. (1990). Fahrzeugführung. In K. Graf Hoyos (Ed.), *Enzyklopädie der Psychologie* (pp. 426–454). Göttingen: Ingenieurpsychologie Hogrefe.
Krcmar, H. (2003). *Informationsmanagement*. Berlin: Springer.
Kroenke, D. M. (2008). *Experiencing MIS*. Upper Saddle River: Pearson Prentice Hall.
Kubicek, H., Horst, U., Redder, V., Schmid, U., Schumann, I., Taube, W., & Wagner, H. (1997). www.stadtinfo.de – Ein Leitfaden für die Entwicklung von Stadtinformationen im Internet. Heidelberg: Hüthig Verlag.
Kuhlen, R. (1990). Zum Stand pragmatischer Forschung in der Informationswissenschaft. In J. Herget & R. Kuhlen (Eds.), *Pragmatische Aspekte beim Entwurf und Betrieb von Informationssystemen*. Konstanz: Universitätsverlag.
Kuhlen, R. (1997). Hypertext. In M. Bruder et al. (Eds.), *Grundlagen der praktischen Information und Dokumentation*. KF Saur-Verlag: München.
Kunz, W., & Rittel, H. (1972). *Die Informationswissenschaften. Ihre Ansätze, Probleme, Methoden und ihr Ausbau in der Bundesrepublik Deutschland*. München/Wien: Oldenbourg-Verlag. http://is.uni-sb.de/studium/kunzrittel.html. Accessed 16 Aug 2011.
Laudon, K. C., & Laudon, J. P. (2008). *Essentials of management information systems* (8th ed.). Upper Saddle River: Prentice Hall.
Lee, Y. W., Strong, D. M., Kahn, B., & Wang, R. Y. (2002). AIMQ: A methodology for information quality assessment. *Information Management, 40*(2), 133–146.
Luckhardt, H.-D. (2000). Hypertext – eine erste Orientierung. In I. Harms, & H.-D. Luckhardt (Eds.), *Virtuelles Handbuch Informationswissenschaft*. http://is.uni-sb.de/studium/handbuch/exkurs_hyper.html. Accessed 16 Aug 2011.
Luckhardt, H.-D. (2005). Information retrieval. In I. Harms, & Luckhardt H.-D. (Eds.), *Virtuelles Handbuch Informationswissenschaft*. http://is.uni-sb.de/studium/handbuch/exkurs_hyper.html. Accessed 16 Aug 2011.
Madnick, S. E., Lee, Y. W., Wang, R. Y., & Zhu, H. (2009). Overview and framework for data and information quality research. *Journal of Data and Information Quality, 1*(1), 1–22.
Martin, J. (1976). *Principles of database management*. Englewood Cliffs: Prentice-Hall.
Mettler, T., Rohner, P., & Baacke, L. (2008). Improving data quality of health information systems: A holistic design-oriented approach. *Proceedings of the 16th European conference on information systems*, Galway, Ireland, pp. 1883–1893.
Payne, S. J. (2003). Users' mental models: The very ideas. In J. M. Carroll (Ed.), *HCI models, theories, and frameworks: Towards a multi-disciplinary science* (pp. 135–156). Amsterdam: Elsevier.
Price, R., & Shanks, G. (2005). A semiotic information quality framework: Development and comparative analysis. *Journal of Information Technology, 20*(2), 88–102.
Ramu, J.-P. (2002). Task structure methodology for electronic operational documentation. *Proceedings of the international conference on human–computer interaction in aeronautics*, AAAI Press, Menlo Park, pp. 62–68.

Ramu, J.-P. (2004a). Electronic operational documentation: From contextualisation towards communication. *Proceedings of the HCI-Aero. Workshop electronic documentation: towards the next generation*. EURISCO International, Toulouse.
Ramu, J.-P. (2004b). *Contextualised operational documentation: An application study in aviation*. http://sunsite.informatik.rwth-aachen.de/Publications/CEUR-WS/Vol114/paper10.pdf. Accessed 16 Aug 2011.
Rossmeier, M., Grabsch, H. P., & Rimini-Döring, M. (2005). Blind flight: Do auditory lane departure warnings attract attention or actually guide action? *Proceedings of the 11th international conference on auditory displays (ICAD)*, ICAD, Limerick, pp. 43–48.
Rumar, K. (1990). The basic driver error: Late detection. *Ergonomics, 33*, 1281–1290.
Schulmeister, R. (2002). *Grundlagen hypermedialer Lernsysteme. Theorie - Didaktik – Design*. München: Oldenbourg Wissenschaftsverlag.
Sklar, A. E., & Sarter, N. B. (1999). Good vibrations: Tactile feedback in support of attention allocation and human-automation coordination in event-driven domains. *Human Factors, 41*, 543–552.
Stix, G. (1991). Computer-Piloten. Trends in der Verkehrstechnik. *Spektrum der Wissenschaft, 9*, 68–73.
Strong, D., Lee, Y., & Wang, R. (1997). Data quality in contexat. *Communications of the ACM, 40* (5), 103–110.
Suchman, L. A. (1987). *Plans and situated actions*. Cambridge: Cambridge University Press.
Tergan, S. O. (1995). Hypertext und Hypermedia. Konzepte, Lernmöglichkeiten, Lernprobleme. In L. J. Issing & P. Klimsa (Eds.), *Information und Lernen mit Multimedia Beltz*. Weinheim: Psychologie Verlagsunion.
Teuber, R. A. (1999). Organisations- und Informationsgestaltung. Theoretische Grundlagen und integrierte Methoden. Wiesbaden.
van Heijst, G. (1995). *The role of ontologies in knowledge engineering*. Amsterdam: Universität Amsterdam, Fakultät der Psychologie.
Verwey, W., & Janssen, W. H. (1988). Route following and driving performance with in-car route guidance systems. IZF-Report No C-14. Soesterberg: TNO Institute for Perception.
Vilimek, R. (2007). *Gestaltungsaspekte multimodaler Interaktion im Fahrzeug*. Düsseldorf: VDI Verlag.
Wang, R. Y., Reddy, M. P., & Kon, H. B. (1995). Toward quality data: An attribute-based approach. *Decision Support Systems, 5*, 349–372.
Wang, R. Y., & Strong, D. M. (1996). Beyond accuracy: What data quality means to data consumers. *Journal of Management Information Systems, 12*(4), 5–34.
Wierwille, W. W. (1993). Visual and manual demands of in-car controls and displays. In R. Peacock (Ed.), *Automotive ergonomics* (pp. 299–320). London: Taylor & Francis.
Wood, S., Cox, R., & Cheng, P. (2006). Attention design: Eight issues to consider. *Computers in Human Behaviour, 22*, 588–602.
Zimmer, A. (1998). *Anwendbarkeit taktiler Reize für die Informationsverarbeitung*. Forschungsbericht des Lehrstuhls für Psychologie II, Universität Regensburg.
Zimmer, A. (2001). Wie intelligent kann/muss ein Auto sein? Anmerkungen aus ingenieurpsychologischer Sicht. In T. Jürgensohn & K. P. Timpe (Eds.), *Anwendbarkeit taktiler Reize für die Informationsverarbeitung* (pp. 39–55). Forschungsbericht des Lehrstuhls für Psychologie II, Universität Regensburg.
Zmud, R. (1978). Concepts, theories and techniques: An empirical investigation of the dimensionability of the concept of information. *Decision Design, 9*(2), 187–195.
Zumbach, J., & Rapp, A. (2001). Wissenserwerb mit Hypermedien: Kognitive Prozesse der Informationsverarbeitung. *Osnabrücker Beiträge zur Sprachtheorie, 63*, 27–44.

Information Ergonomics

Heiner Bubb

1 Categorizing and Definitions

1.1 Micro and Macro Ergonomics

"Ergonomics" is a multi-disciplinary science which uses basic knowledge from Human Science, Engineering Science, and Economic and Social Science. It does comprise occupational medicine, industrial psychology, industrial pedagogic, working-technique and industrial law as well as industrial sociology. All of these sciences, from their different points of view, deal with human work and therefore represent an aspect of this science. With regard to feasibility, this basic knowledge is summarized in so-called praxeologies (see Fig. 1). The more economic-academically and social-academically oriented of them is the "macro ergonomics", which provides rules for the creation of organization, company groups and study groups. The more engineering oriented "micro ergonomics" gives rules for the technical design of workplaces and working means. The latter refers at first in narrower sense to the interaction between person and machine. However, in both cases the special focus of the research is directed upon the individual person and his experience of the situation in the workplace what represents the ergonomic specialty compared with the other science disciplines which argue with similar contents.

The central object of the ergonomics is to improve by analysis of tasks, of environment and of human–machine-interaction the efficiency of the whole working system as well as to the decrease of the charges having an effect on the working person (Schmidtke 1993).

H. Bubb (✉)
Lehrstuhl für Ergonomie, Technische Universität München, München, Germany
e-mail: bubb@tum.de

M. Stein and P. Sandl (eds.), *Information Ergonomics*,
DOI 10.1007/978-3-642-25841-1_2, © Springer-Verlag Berlin Heidelberg 2012

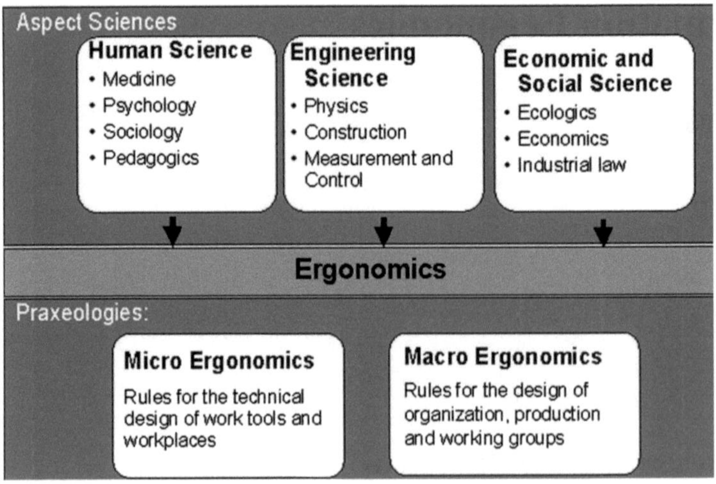

Fig. 1 Ergonomics and their disciplines

1.2 Application Fields of Ergonomics

Today it has established as meaningful to divide also ergonomics after its areas of application. Thus it can be distinguished especially between product ergonomics and production ergonomics:

- *Product ergonomics*
  The priority aim of product ergonomics is to offer a very user-friendly utensil for an in principle unknown user. Therefore, for the development of such products it is important to know the variability of the users with respect on their anthropometric properties as well as their cognitive qualities and to consider this during the designing process. An actual and new field of research of the product ergonomics is the scientific question what is the meaning of *comfort feeling*.
- *Production ergonomics*
  The matter of production ergonomics is to create human adapted working places in production and service companies. Here is the aim to reduce the load of the employee and to optimize at the same time the performance. In contrast to the case of product ergonomics the employees are ordinarily known and it can and must be considered individually their needs. According to all experience the acceptance of ergonomic measures is only guaranteed in production processes if they are carried out in co-operation with the employee.

As the above described methods of the ergonomics are applied as with product as with production ergonomics and because often the "product" of one manufacturer is "working means" of the other, an exact separation is not possible between these both areas of application practically, particularly as in both areas the same scientific methods and conversion procedures are used.

Areas of application of primary importance, in which today systematically ergonomic development is operated, are the range of aviation (special cockpit design of the airplanes, design of the radar controller jobs), the vehicle design (passenger car and truck: Cockpit design; anthropometric design of the interiors, so-called packaging; design of new information means, by which security, comfort and individual mobility are to be improved), design of control rooms (chemical plants, power stations; here above all aspects of the human reliability play an important role) and offices (organization of screens, office chairs, the total arrangement of workstations, software ergonomics). Also the development of well understandable manuals can be assigned to the area of ergonomics. A further special field of ergonomics is the research of limit values for working under extreme conditions like extreme spatial tightness, cold weather, heat, excess pressure, extreme accelerations, weightlessness, disaster control operation, etc.

2 Object of Ergonomic Research

2.1 *What Is Work?*

If within the scope of ergonomic design the work and the working means – it is the machine – should be designed appropriate for human properties it is to be answered first the question what should be understood by work in this connection. Work and work content are certainly determined on the one hand by the working task, on the other hand, however, also essentially by the qualities of the tools which are developed for the increase of the working efficiency or to the relief of the worker. Work is divided traditionally in physical and mental work. The technical development allows expecting that furthermore a substitution of physical work by suitable machines takes place and with it a far-reaching shift of physical work to mental work.

With view on manual work the investigations of Taylor and his co-worker Gilbreth play a basic role. The systems of predetermined times are based on their works which dismantle manual works in activity elements and assign to these times as a function of the accompanying circumstances. In general more than 85% of all activities can be described by the working cycle consisting of the activity elements "reach", "grasp", "move", "position", "fit" and "release" (see Fig. 2).

As can be shown, this work beside the physical aspect of the "Move" has essentially information changing character (Bubb 1987). Already in 1936 Turing has carried out a general description of such processes of information changing (see Fig. 3): The person respectively the information processing system perceives a symbol, is able in the next step to extinguish this symbol respectively to headline it and shifts now in the third step the neighbouring, in its arrangement necessarily not recognized symbol to the place of the original symbol. This process marks an ordering process. It is easy to understand that the activity elements "reach", "grasp" and "move" can

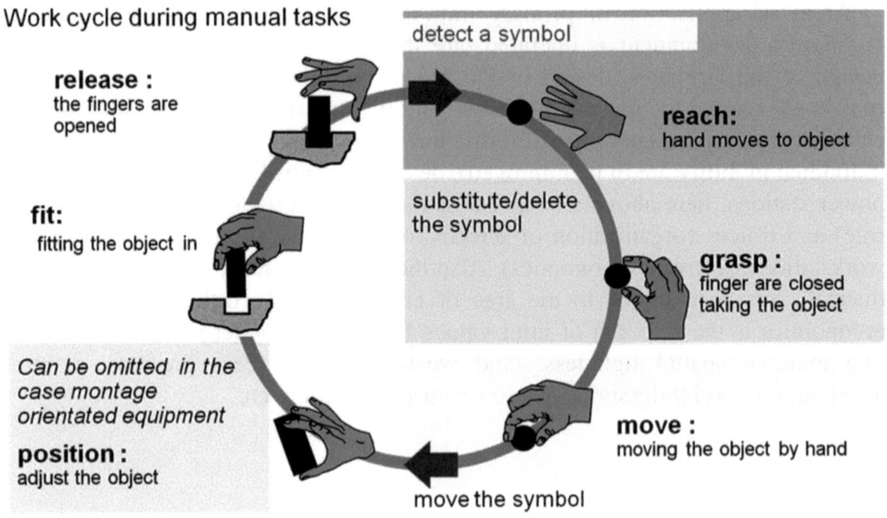

Fig. 2 Motion cycle in connection with manual activities

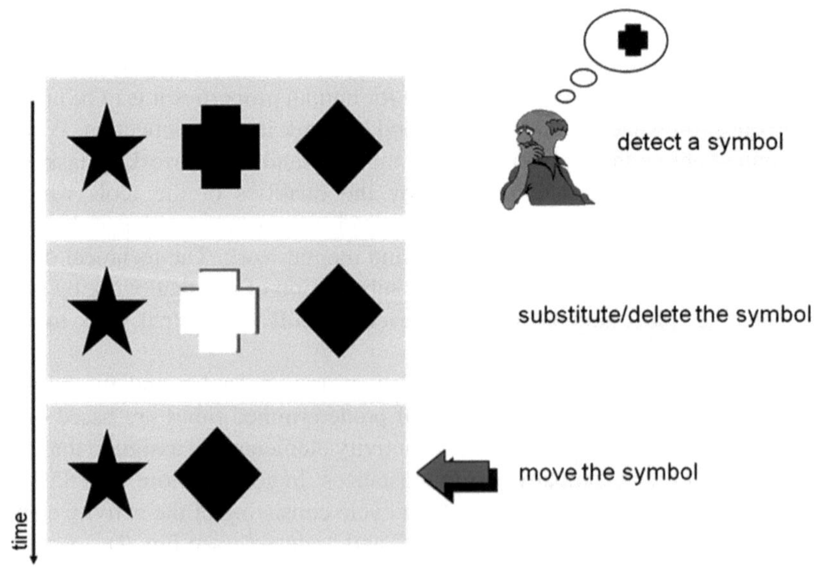

Fig. 3 The Turing machine as a model of thinking processes – elementary basics of ordering

be assigned to these three Turing elements. The ordering process of an assembly is closed with these three activity elements. The further activity elements are obviously only necessary to fix the new order. They can be partially or also completely cancelled by a favorable "rational" design of the working place (e.g., "position").

Also pure mental activities are to be seen under this aspect of "ordering". Also in this case it is principally possible to separate the substantially order producing elements of work from such that have only fixing or preparing character. An essential aspect of ergonomic design consists in recognizing unnecessary working steps and in removing or in simplifying it at least by technical design.

In generalization of the considerations shown here and according to the scientific concept of information after which information is a measure of the divergence of the probable autonomously, i.e. "naturally" appearing state (information = negentropy), can be concluded that every human work shows an ordering process, by which against the natural trend of the increase of the entropy (= decrease of negentropy or information) an artificial human produced state is created (Bubb 2006). By producing such an ordering energy turnover is always implied for the supply of the necessary movement. This energy turnover can be raised directly by the human activity (physical work) or by a machine, which must be operated, however (mental "operating work").

The here presented scientifically oriented working concept goes out from effects visible to the outside, i.e. from an objective point of view. After that work is production of an artificial human created order, i.e. the production of a work in terms of goods and service (Hilf 1976). The humanities-related working concept characterizes on the other hand the subjective experience of the work (Luczak 1998). This is frequently characterized by "trouble and plague", however, can be also depending on the internal settings a cause for an increase of self-esteem and subjectively positively experienced satisfaction. For a comprehensive working analysis both aspects are to be considered. The demand of order production characterizes the (in principle objectively detectable) load, the subjective experience the strain depending on individual qualities, abilities and settings.

Basis of every Man Machine System (MMS) design is therefore the flow of information which conveys the task that should be accomplished with the system transfers in the result. This human machine interaction can be illustrated especially well by means of the metaphor of the control circuit: the human operator transforms the information given by the task into a control element position which is conveyed by the working principle of the machine in the output information, the result. Normally the possibility consists to compare task and result with each other and to intervene with a not accepted deviation correcting in the process. This process can be influenced by circumstances which deal nothing with this real task-related flow of information. They are designated in this context "environment" or "environmental factors" (see Fig. 4). The degree of the relationship of the result R and the task T represents the working quality Q:

$$Q = \frac{R}{T} \tag{1}$$

The working performance P is the working quality achieved per time t (Schmidtke 1993):

$$P = \frac{Q}{t} \tag{2}$$

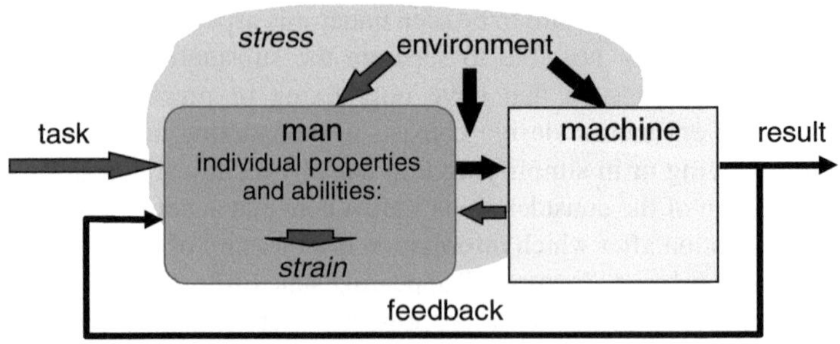

Fig. 4 Structural scheme of human work

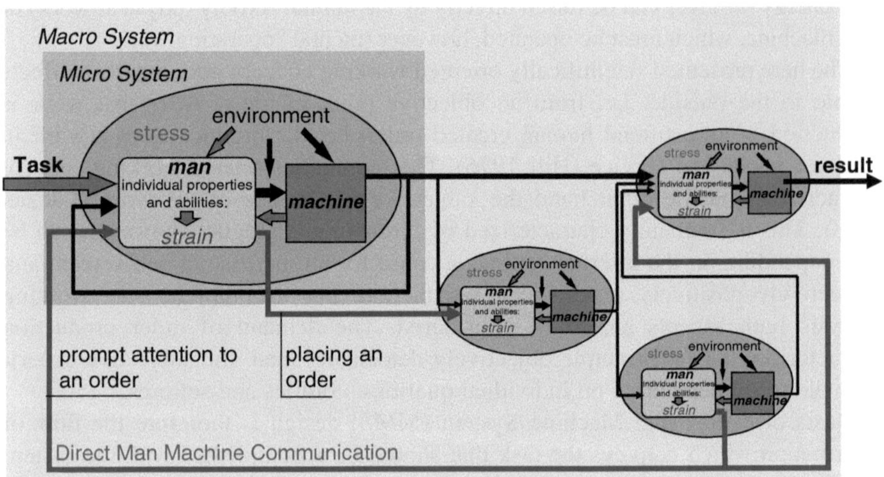

Fig. 5 Information flow in a macro system consisting of several single human machine systems

The aim of ergonomic design is to reach an optimization of the performance by an adaptation of the working means (= machine) and the working environment to the human operator.

If one extends the Fig. 4 by the possibility of the operator to prompt attention to an order and to place an order to other, the micro system of the simple human machine system can be extended to the macro system of complicated equipment (Fig. 5).

In any case, the individual human performance during accomplishing his work is influenced by the so-called external performance shaping factors and by the so-called internal performance shaping factors (see Fig. 6). It is a job of the management to create the requirements for an optimum performance application of the operator by the design of the external conditions. Thus the connection is produced between the ranges of Micro Ergonomics and Macro Ergonomics.

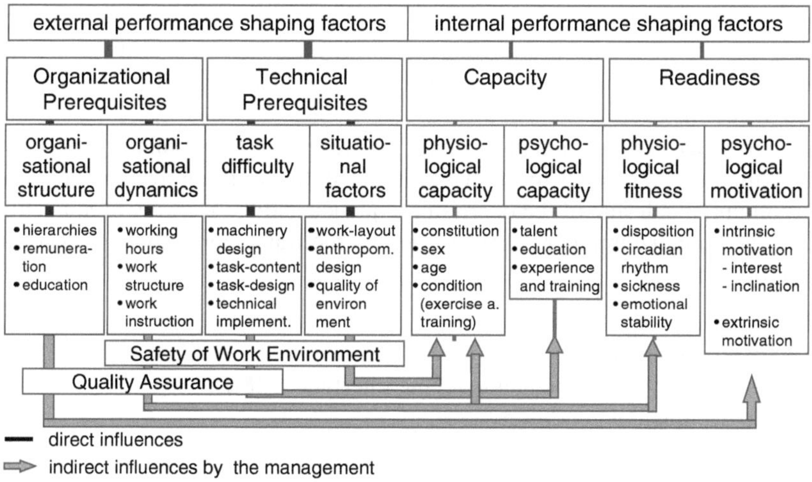

Fig. 6 The working performance influencing preconditions

3 The Position of Information Ergonomics in the Field of the Science of Work

3.1 *System Ergonomics*

From the preceding discussion the basic approach becomes visible to understand every working process as an information change. As the operator is integrated in a – if necessary complicated – system, the comprehensive consideration of the total system is of prominent meaning for the design of the work appropriate for human worker. This is designated in general *system ergonomics* (Hoyos 1974; Döring 1986; Luczak 1998; Kraiss 1989; Bubb 2003a,b). Within the scope of the system ergonomics the information flow is investigated that result from the integration of the human operator in the man machine system. The design produces the optimization of this information flow. In order to achieve it two destination directions are of importance that excludes them mutually by no means. In order to reach the aim in both cases often the same measures can be used: On the one hand, the system can be optimized for the purposes of the working performance as defined above and on the other hand for the purposes of the system reliability. In the first case it is a matter of raising the effectiveness and the accuracy of the system, i.e. the quality achieved per time unit. The applied methods are taken over largely from the control theory. Essential consideration objects are the dynamic properties of the system for whose description particularly transitional function and frequency response are suited. In the second case it is the aim to receive the operational safety of the system or even to raise it, i.e. to minimize the probability for the fact that the system gets out of control. The applied methods are those of the probabilistic. In order to describe the interaction of the total system, the methods of the Boolean algebra are applied. For the appearance of failure probabilities or from mistakes probability distributions must be assumed. In particular the Human Error Probability (HEP) plays a role in this connection.

Design ranges of the human machine interface and the system ergonomics are related among other things to the working place, the working means, the working environment, the design of the operating sequence and the working organization as well as the human computer interface. Hereby the frame conditions (of design) are among other things:

- The working task,
- The external conditions,
- The technology of the human machine interface (MMI),
- The environmental conditions,
- Budget, and
- User's population (Kraiss 1989).

In the focus of the mentioned aims based on Schmidtke (1989) the following objects are to be mentioned:

- Functionality (adequate support of the coping of the problems),
- Usability (intelligibility for potential users),
- Reliability (error robustness, that is "less" operating errors are tolerated),
- Working safety (avoidance of health damages),
- Favorability of individuals (consideration and extension of the user's ability level).

Further aims are economic efficiency and environmental compatibility (Kraiss 1989).

With regard to the information change within the system – especially concerning the capturing of the task – one can make following differentiation from the point of the integrated individuals:

- The information of the task and in many cases also the result is received from the natural environment. This is, e.g., the case with the steering of land vehicles, where the necessary reactions are taken up visually from the natural environment that contains the traffic situation. The same is also valid for operating quick vessels and partly for operating airplanes during the start and in particular during the land approach. Another example is observing the executions and the action by workers in the assembly line. Many working processes – for example, in the craft range – are characterized by such "natural" information transmission.
- The information of the task or partial aspects of the task, the whole fulfillment of the task or certain measurable working conditions of the machines are given artificially by displays. In this case the optimization of the process refers not only to the interaction possibilities and the dynamics of the system, but substantially also to the display design that should be realized in such a way that contents and meanings of the processed information are reasonable for the person very

immediately. Because here the focus lies on the design of this information processed by a technical way, this special branch of the system ergonomics is designated often also *information ergonomics* (see below; Stein 2008). In this connection is to be mentioned the so-called "Augmented Reality (AR) technology" by which artificial generated information is superimposed to the natural environment. Different technologies are available for this. Fix mounted Head Up displays (HUD) in vehicles and with the head of the operator solidly connected Helmet Mounted Displays (HMD) are most common today. But also the methods of the holography or special laser technologies are discussed in this connection. By means of these optical principles it is possible for the operator to receive at the same time the natural environment and the artificially generated information.

The increasing mediatizing and informatizing given by the technical development let increase the importance of information ergonomics, as under the condition of the complex technology it becomes more and more necessary to provide the user with quick, unequivocal, and immediately understandable information (Stein 2008). Especially the safety demands in aviation offer a big challenge for application of information ergonomics.

3.2 Software Ergonomics

The professional life was changed very comprehensively by the introduction of the computer technology. If only rather easy logical linkages were possible with the traditional technology (e.g., mechanical realization of the basic arithmetical operations by means of engine-operated or hand-operated calculating machines respectively by means of the relay technology realized simple representations of the Boolean algebra in electric circuits), so by the computer technology demanding "mental processes" can be also copied now, decisions are made, complicated graphic representations are generated in real-time and much more. The similarity of computational information processing to that of the human operator, but also the indispensable stringent behavior of the computer pose quite specific problems to the user. The adaptation of program flows, of interventions in this, as well as the suitable representation of the results for the user has become an own branch of the ergonomics which is assigned by "*software ergonomics*". By it is understood the creation, analysis and evaluation of interactive computer systems (Luczak 1998; see also Volpert 1993; Zeidler and Zellner 1994). With the software ergonomics there stands the "human appropriate" design of user interfaces of a working system (among other things dialog technology, information representation), considering the task to be performed and the organizational context in the centre of the ergonomic interest (Bullinger 1990; Bullinger et al. 1987; Stein 2008).

On account of its meaning software-ergonomic demands are agreed effectually in European, as well as national laws, as well as partly on voluntariness to based standards. Beside different more generally held guidelines particularly the *DIN EN ISO 9241* – 110 and *DIN EN ISO 9241* – 303 is to be mentioned here.

Particularly the principles of the dialog design are highlighted here:

- *Task suitability*: A dialog is task appropriate if it supports the work of the user without unnecessary strain by the attributes of the dialog system.
- *Self-describability*: A dialog is self-descriptive if the user will be informed on demand about the scope of performance of the dialog system and if each single dialog step is direct intelligible.
- *Controllability*: A dialog is controllable if the user manages the dialog system and not the system the user.
- *Consistent with expectancy*: A dialog is functional if it does what the user expects out of his experience with previous work flow or user training.
- *Fault tolerance*: A dialog is fault-tolerant if in spite of faulty entries the intended working result can be reached with minimal or without correction.
- *Can be customized*: A dialog should be able to be adapted to the individual needs or preferences of the user.
- *Learnability*: A dialog i.e. easy to learn if the user is able to manipulate the system without reading help functions or documents.

3.3 Information Ergonomics

The technical development has led to the interlinking of the computers, assigned in the companies internal as "Intranet" and worldwide as "Internet". In these networks in a unconceivable manner information is transported, processed and made available from the point of view of the user's as good as detached from a material supporter (of course, if this is not correct actually!). The hypermedia organization of the information, the investigation of human search strategies, whose result is a requirement for the creation of search engines and again the technical representation of the so gained information, is a main object of the already mentioned *information ergonomics*. The concept information ergonomics is often shortened related to the creation of Internet-based information systems, here in special the navigation design. In the foreground here stands the support of search strategies (searching) which should lead fast to the required information. In addition, beside the support of searching by adequate search tools a support of the "rummaging" or "going with the tide" is also aimed in the cyberspace (Browsing; see Marchionini and Shneiderman 1988). This means that information systems must be designed in such a way that different navigation patterns and therefore not only purposeful actions are supported (see also Herczeg 1994). Consequently not only the not planned, but also the implied and interest-leaded action is considered. Thus the search is often based on a voluntary interest in information which can be used partly for professional as well as for private purposes. Competing information and theme areas are only one "click" nearest of each other. This requires, by means of media design to create an incentive (attraction) and to hold this for the period of use in order to avoid e.g., psychic tiredness, monotony and saturation and to allow states like "flow".

After Stein and Müller (2003a, b) the goal of information ergonomics is the optimization of information retrieval which includes information search and selection (appraisal and decision making processes) as well as cognitive processing of text-, image-, sound- and video-based information of information systems. In this respect, information ergonomics comprises the analysis, evaluation and design of information systems, including the psychological components of the users (capabilities and preferences, etc.) as well as work tasks (information tasks) and working conditions (system and environmental conditions) with particular emphasis on the aspects of motivation, emotion and cognitive structure (content knowledge and media literacy). User motivation and emotion during information system access and retrieval processes are fields of investigation and analysis (Stein 2008).

On the part of the information system information retrieval is affected by the implemented information structure (organization and layout) and by the design and positioning of navigation aids as well as by information contents and their design. The exactness and swiftness of perception, and selection (appraisal and decision making) of information with regard to their pertinence to professional or private needs (information) on the part of the recipients primarily depends on text, image, sound, video design and structuring. This also applies to the quality of cognitive processing.

Information structure, content categories, navigation instruments, search functions, graphics/layout, links, text/language, image, sound and video design etc. represent levels of design. Design should enable the user to successfully perform information retrieval, appraisal and decision making, selection and cognitive processing within the scope of a work task or a self-selected goal within a reasonable period of time. In addition, the information system should be designed to enable the user to exceed his initial skill level with regard to his media literacy and expertise (Stein 2008).

Figure 7 shows the general structure of information ergonomics: It is the task of information system to transform the information system access into the application of information to a control task. The information system access is essentially influenced by the system, the task and the environmental conditions. These conditions define the necessary information retrieval. In order to design the information system and with it the information retrieval according to ergonomic demands the user with his information goals, his expectations, motivation, emotion knowledge and media literacy is to be considered essentially.

Thus seen information ergonomics represents a specialization of system ergonomics whereas especially such systems or system parts are considered in which the technical processing of the information dominates.

3.4 Communication Ergonomics

Whereas the information ergonomics refers partly substantially to the use of the Internet, the use of the Intranet belongs to the so-called *communication ergonomics*. Concerning this it must be highlighted that user's interfaces differ in the production area generally considerably from such in information systems. While Internet-based information systems are based as a rule on information tied up to links (hypermedia) which opens for their part a huge number of further information, by contrast as a rule user interfaces in the production area allow access on a high detailed, but limited thematic area. After Zülch (2001a) object of the communication ergonomics is to design the information exchange between human operator and technical system in such a way that the abilities, skills and needs of the operator are considered appropriately whereby the aim is to design user friendly economic courses, products, jobs and working environments. Hereby especially mental overloading or under loading should be avoided to achieve the best possible result of working (Zülch 2001a; Zülch Stowasser and Fischer 1999; Kahlisch 1998; Simon and Springer 1997). The communication ergonomics contain the "creation of use interfaces in the production range, evaluation of CIM interfaces, navigation in object-oriented data stores, creation of multimedia documents, and standard conformity tests" (Zülch 2001b). Concerning this all working places with screens are an object of the ergonomic interest.

3.5 Cognition Ergonomics

In the before mentioned areas of ergonomic research and application it is primarily about the information flow and less about the physical design of the working place. Information processing by the human operator makes demands for his cognitive abilities. That is the reason why the concept *"cognition ergonomics"* has also recently become the custom as a name for the areas described above.

3.6 Application Fields

The system-ergonomic design, respectively the application of information ergonomics is necessary because of the penetration of all technology with computer technology today in almost all scopes of work. Many devices of the daily life (remote control for electronic entertainment devices, electric washing machine, the entertainment system in cars, etc.) as well as electric tools, machine tools, like NC- and CNC-machines, robots in the production area, but in future also more and more in the household area are examples for this. In the context of company internal information systems a specific information management is necessary. A special attention is necessary here for the knowledge loss which is given by retired employees. It is expected to absorb this loss with computer based knowledge management systems. The progressing integration of the domestic entertainment media on the Internet makes necessary user-friendly information systems. Generally the Internet revolutionizes the accessibility and use of information and of libraries. In the supervision control centers for traffic engineering (e.g., traffic engineering for the urban traffic, air traffic controller, shipping pilot centrals), in the control centers of power plants, and chemical equipments the access to special computer programs occurs. The passengers of public transportation systems (in particular railway and air traffic) receive information about timetables as well as the tickets from computer-controlled machines. The control of the vehicles is supported more and more by so-called assistance systems. If for a long time the aviation has played here forerunners roll, such systems penetrate today more and more into the area of the privately used motor vehicles.

The enumeration of these application areas makes evident a difference between the users. While one can be compensate in the commercial range lacking ergonomic design of the interaction by suitable training and training programs, one must go out in the area of private use from the untrained and unskilled operator. There the highest ergonomic demands are made. The application of the rules which one achieves in this area will also bring in the commercial area advantages in form of more reliable and more efficient operation. Not at least this will lead to long-term reduction of costs.

4 The Object of Information Ergonomics

4.1 System Technological Basics

In general it can be ascertained: The application of system-theoretical principles to ergonomic questions is assigned by *"system ergonomics"*. The system theory itself has its origin in the information transfer technology (Wunsch 1985). It treats every considered complex of (natural) phenomena in principle according to the same basic scheme. The advantage of it is to be able to treat even the human operator in his interrelation with the technical environment with uniform methods.

By *system* is understood the totality of the necessary technical, organizational and/or other means that are necessary for the self paced fulfillment of a task complex. Every system can be thought composed of elements which interact to each other. The essential aspect of the system theory lies in the fact that the physical nature of the interaction is ignored (e.g., force, electric voltage/current, temperature, air or body sound, etc.) and only the information contained in it respectively the information change caused by the system is considered. In this connection by information is understood every deviation of the natural distribution of energy and matter, as already shown above. Practically the aimed information change can consist in the change of matter (task of the "machines") as well as in the change of energy (task of the "prime movers" or "energy production") as well as in the spatial change of energy/matter (task of the hoisting devices, conveying plants, pumps, fans, vehicles) or in the change of matter/energy distributions (task of the information-processing machines, as for example computer, data carrier, and so on). During his operation the human operator transmits information on the machine and provides in such a way for the fact that this manages the effects desired from him.

The desired change can only occur, as the system stands in information exchange with the so-called system environment (so-called "open system"). The system has therefore an input side ("input") by which from the environment information (i.e. energy or matter) on one or several channels is taken up and a output side ("output") by which the changed information on one or several channels is delivered. All effects which influence the desired process of the information change by the system are designated in this sense by "environment".

In order to get a more exact idea of the functions and effect mechanisms of the system, it is thought composed of *elements* and these elements stand with each other in the information exchange. The active relations of the elements are described by the *system structure*. The canals of information are shown graphically as arrows which mark therefore also the active direction of the information. Within such an effect structure the parts are called the elements which change the detailed information in typical manner. One distinguishes elements which connect information with each other or such elements which information distribute (convergence point, branching point, switch), and elements which change information according to a function and again those are separated, with which the time plays no role ("elements without memory"), from such with which the time plays a role ("elements with memory"). Elements with memory cause a system dynamic. As the human operator is always an "element with memory" in any case a human–machine-system has a dynamic that means its behavior is time dependent.

Under the shown aspect the machine as well as the human operator can be seen as a component of a *"human machine system"*. One can distinguish basically two manners of connection: A *serial connection* describes the case that the person takes up the information of the task, converts it in adequate manner and transfers it by the control elements on the machine. This changes – mostly under change of separately added energy – the input information in the deliberate result (so-called *active system*; e.g., car driving: the task lies in the design and lay-out of a road and in the situation presenting there, the result is the actual position of the vehicle on the street). Generally the operator can observe the result and derive new interventions by the control elements from the comparison with the task. It is a closed-loop *controlled* process. If, nevertheless, the time becomes too big between the intervention by the control element and the recognition of the result, he must derive only the control element activity from the task. Then one speaks of an open loop *control*. The process can be influenced by environmental factors. In this case more exactly is to be fixed, whether this influence refers on the human operator and his abilities (e.g., noise), on the transfer between human actor and machine (e.g. mechanical vibration) and on the machine itself (e.g. disturbance of the function by electromagnetic fields).

In the case of *parallel connection* of human operator and machine an observer's activity comes up to the operator (so-called *monitive system*). The machine, designed as an automatic machine, changes independently the information describing the task in the desired result. If the person observes inadmissible divergences or other irregularities, he intervenes in the process and interrupts the process or takes over the regulation from hand. Also this process can be influenced by the environment, whereby again the operator, the machine and the transference ways are the typical influencing points.

The system development is to be divided into temporal *phases* and duties typical for phase are to be done. Basis for it is the *life cycle of* a system. As Döring (1982) has worked out, the role of the human operator is to be considered in the *concept and definition phase* in an order analysis and function analysis and the functional subdivision arising from it between person and machine. Then in the *development* and *production phase* the exact design of the interface between operator and machine occurs, while in the *installation and operating phase* the final assessment takes place. For the *decommissioning phase* the same cycle is to be considered like for the commissioning phase. Its meaning recently becomes evident, actually, when one only recognizes, that the removal of the so-called industrial garbage cannot leave any more only to scrap metal traders.

4.1.1 Application of the "Cause Effect Principle"

If one describes the function of elements within a system by functions in principle on the base of a given input function an unambiguously output function or result function can be predicted. This thought corresponds to the usual *cause effect principle*. The basic advantage of this method on which practically the success of the whole technical development is based lies substantially in the fact that relatively long-term predictions are possible and that in complicated systems the temporal-logical structure of the interaction of the single elements can be fixed precisely.

Aim of the closed loop control circuit is to follow time depending forcing function and/or to compensate time depending changing disturbance variables. With it the dynamic qualities of the single system elements and the whole control circuit play a dominant role. In order to describe these dynamic qualities or to test, two procedures more differently complementary to each other are used, by which two extremes of the behavior of the system elements or the whole control circuit are described: by the so-called *step response* the reaction to a suddenly appearing change (e.g. in the case of car driving: Interception of the car with a suddenly appearing wind gust) and by the *frequency response* the behavior in the so-called steady state (e.g. in the case of car driving: To continue along a bendy country road) is described. With both procedures the basic idea of the quality is used to the mathematical description: whether related to the whole control circuit or related to a single system element, always the relation of the output signal to input signal is formed. In the case of the step response these dimensions are looked as a function of the time and in the case of the frequency response as a function of the frequency.

In most cases it is difficult – if not impossibly – to describe the behavior of the human operator for the purposes of mathematical functions. Nevertheless this happens in special cases, e.g., in the form of the so-called "paper pilot": In this case the frequency response of the human operator is formulated mathematically. This modeling permits to a certain extent the advance calculation of the effect of different technical qualities of the machine – especially of an aircraft – on the system behavior in connection with the controlling person (see also see also McRuer and Krendel 1974). However, this modeling can have validity in this form only for *highly skilled activities*. In the situations which demand rather cognitive abilities of the person, the modeling which describes his *decision behavior* is necessary.

4.1.2 Probabilistic Methods

Now, however, the experience shows that human and technical system elements do not function under all circumstances for the thought purposes. With respect to the above referred image, that also the whole system shows an order deviating from the "natural" circumstances – that means that it represents information – is to be expected that its structure disintegrates by itself according to the second law of the thermodynamics, and so his function loses. This is the cause for the failure of elements and systems, however, also on the side of the human organism of the forming of illnesses and inadequacy. In the case of human operator as an adaptive system additionally can be assumed that he is possibly almost enforced by an "internal deviation generator" to make occasionally mistakes in order to develop new experience from it. Of course in the technical area the decay speed is very different as a function of specific material conditions (see, e.g.: Information of a writing in stone chiseled or in sand written); the decay speed can be influenced by choice of the material, the thickness of the material or the cleanness of the material to a great extent (nevertheless, the object of the traditional security technology), however, the decay cannot become completely prevented. The inherent randomness of the decay marks also the meaning of bringing in human action for servicing and maintenance tasks: while in clear cases it still succeeds to automate the (regular) normal process because here – indeed, only in less complicated cases – the possible influence to taking conditions can be thought ahead, this is practically impossible in case of the servicing and especially maintenance tasks. The application of the human operator is necessary here who can react by means of his creative intelligence to the different, in principle no predictable cases.

In order to define the probability of the appearance in the case of the countable error is to be fixed under which conditions one speaks of an error. The definition 1 of the quality can be taken again for it. The quality defined as analogous (i.e. continuous) value is usually digitized in the practical use, while quality classes are formed. The easiest form shows the classification "quality o.k." – "quality not o.k.". Then an accomplishing of the task beyond demanded quality borders, i.e. beyond the *"acceptance area"* will be called *"error"*. This error concept stands according to the error concept of DIN V 19 250 (1989) which understands under this the deviation going out a defined range of tolerance of a unit. If this error is caused by a human action, this is called Human Error.

One counts now the errors afterwards (*a posteriori*) and forms their relative frequency or determinates from the start *(a priori*) on the basis of mathematical calculations and assumptions of the probability distribution the likelihood for the appearance of an error. By this one receives the numerical values with which one can calculate the error probability of single elements and also of the human operator. So, by use of the Boolean algebra the total error probability can be calculated including the human influence.

The way of thinking in probability (so-called probabilistic procedure) shown only sketchy here may not be understood as an alternative to the classical cause effect way of thinking. The function of a system can be achieved only with this and formed by means of scientific methods. However, in addition, the probability oriented approach delivers other possibilities to gain control of the reliability and security of an increasingly more complicated becoming technology and in particular also the mashing of this with the environment and the controlling person.

4.2 Information Technological Human Model

4.2.1 Human Information Processing

In order to describe the interaction between human operator and machine with the aim of the design according to an engineering approach, the technical information modeling of the human operator is necessary. The system element man is represented graphically like every system element as a rectangle with an input (ordinarily on the left side) and an output (ordinarily on the right side). The input side of the person, the *information perception* is defined by his senses, like eyes, ears, etc. The information taken up is changed in transformed form on the output, the *motor processing* into one of the outside world sensible information. This is characterized by the possibility to bring the extremities hands and legs in desired positions (and motion), but also by the speech. The central nervous system which connects together the from the outside coming information with each other, carries out logical derivations, and makes decisions, is called *information processing*.

However, with this characterizing is to be considered that a sharp separation is possible between these human subsystems neither from physiological nor from psychological view. In the information processing in the narrower sense the taken up information is connected together with commemorative contents by logical deriving, drawing of conclusions and by decisions. Rasmussen (1987) describes three level of information processing:

1. Often repeated connections which turn out useful are learned as combined actions (so-called internal models) which can be realized unconsciously, virtually automatically, i.e. with low (or to none) deliberate expenditure (*skill-based*

behavior). These actions are quick: The time difference between the stimulus and the reaction lies between 100 and 200 ms (e.g., driving a car on a winding street).
2. Furthermore taken up signals can be processed by reminding and the use of stored rules, after the pattern: "if… a symptom…, then… an action" (*rule-based behavior*). This action manner requires a certain deliberate expenditure (e.g.: Follow the "right has priority for left" rule in the traffic). The searching and the choice of these rules require the expenditure of seconds or even minutes depending on the different situations.
3. Planning, decision and problem solving in new situations are mental achievements at the highest level of the information processing and correspond to the so-called *knowledge-based behavior* (e.g., diagnosis of the failure causes of a complicated system). According to kind of the task and the measure of the closeness linked with it the treatment of parallel tasks is not possible. These actions require as a function of the difficulty of the task a time demand of minutes to even hours.

In general the probability of a false action rises if the treatment time necessary at the respective treatment level ("required time") is given equal or is even greater than objective the time span within which the task is to be done ("available time"). For a significant degradation of the error probability the available time must be bigger possibly around the factor 2 than the required time. In order to guarantee quick perception of technically displayed information the aspects of information retrieval as shortly shown in Fig. 7 are of outstanding importance.

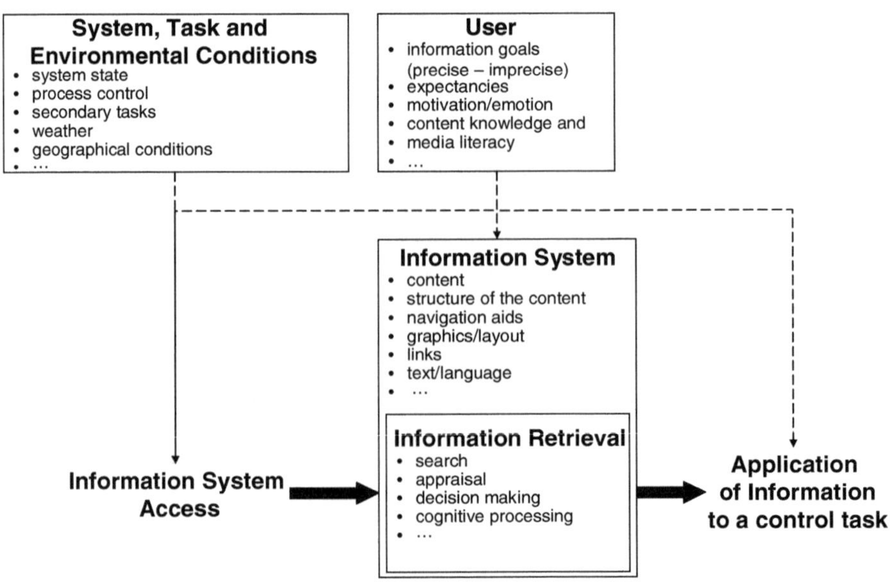

Fig. 7 Control circuit of the human machine system with the nomenclature usual in the control engineering

For the design of machines are the works of Pöppel (2000) of great importance. He found out that the area of the immediately experienced presence is given by about 2–3 s. In particular many investigations from the traffic research support the image of this limit which is given by this 2–3 s horizon (e.g., Gengenbach 1997; Crossmann and Szostak 1969; Zwahlen 1988). However, Pöppel also points to the fact, that additional to this immediate present feeling still the area of the so-called "present of the past", i.e. the recollection of the immediately before events and the so-called "present of the future" belongs. The last refers to the expectation of that what happens at the next moment. The whole span of present of the past up to present of the future puts out about 10–15 s and encloses the time span, in which the so-called working memory immediately can be processed.

Although every human *motor processing* can be led back basically on muscle movement, must be distinguished in practice between the change of information by:

- Finger, hand, foot activity and
- Linguistic input.

The technical equipment can be adapted to the conditions of the motor processing by suitable design:

- Adaptation of the control elements to the anatomical qualities of the person;
- Sufficient haptical distinguishability of the control elements;
- Enough force feedback by the control element;
- Feedback of the influenced process.

The transference of information by language plays a big role above all with the intercommunication between people. In general communication problems are on the one hand effects of the perception of information (e.g., masking of the language by noises) and on the other hand of semantic kind. To the latter problem is thereby tried to prevent often that one uses standardized expressions within a certain technical system (e.g.: air traffic). However, the linguistic input also plays in the human machine communication increasingly a role. The performance and reliability of the conversion of information by language (linguistic input) is influenced by factors which are due to the speaker, to the used word repertoire and text repertoire, from the qualities of the speech recognition system and to surroundings noises (see Mutschler 1982). In this area are reached at the moment of big progress (Wahlster 2002).

However, in general no human activities exist that are total errorless. In many cases the human operator recognizes often immediately that he failed. In this connection information retrieval plays additionally an important role. By technical means is to be provided that every step that was initiated by the human operator can be retrieved at any time. Often that can be realized by an "undo function" that recalls the last step. In more complicated cases additionally a back step to the start position or a meaningful interim state is necessary.

4.2.2 Modeling of Human Decision Behavior

For the development of technical aid it is basic among other things to identify the dimensions of influence for decisions. The modeling of human decisions serves it. One receives a clear representation in form the so-called decision matrix (Schneeweiß 1967). In this the respective circumstances under which an action occurs are laid down for example in the columns and in the lines the possible actions. The likelihood of the appearance of these situations is estimated on the base of the experience (internal model). Then a (thought) event can be assigned to every combination of situation and action to which in each case an individual benefit or damage (= negative benefit) corresponds. Then the action is selected, which promises most benefit in the sum or allows expecting the slightest damage.

With regard to the technical human models based on control theory as well as the modeling of the human decision is to be bent forwards a wrong claim appearing over and over again. The aim of such ergonomic human models does not consist in "to explaining" the human being. It is not a matter of uncovering the effect mechanisms on biological, physiological or also psychological base precisely for the purposes of a "establishing of the truth" and of reaching with it a progress in the science of the knowledge about the human existence. This stands, for the rest, in perfect analogy to the anthropometric human models which also want to fulfill no medical–physiological claims. The objective is in both cases of rather purely pragmatic nature: it is, to simplify the won knowledge about the human behavior in the different appealed science disciplines and to concentrate it in a model in such a manner that can be derived from its – very unequivocal – conclusions for the technical design. Jürgensohn (2000) found out, for example that control theory based human model can represent always only one limit area of human behavior. Therefore, in particular the rather technically oriented models should be connected in future with models from the psychological area. But also cognition ergonomic human models want to predict the realization of certain decisions neither nor explain, but from a rather general view derive in similar manner tips to technical improvements as it is possible for the dynamic interaction on the fundament of control theory based models.

5 Aspects of Information Ergonomic Design

On the basis of the preceding exposition the cognition ergonomic design area can be founded. Every working task requires a cognitive processing expenditure from the person. Depending on the design of the working means this processing expenditure can be reduced and with it the mental load, but be raised also. The optimization of the mental load, i.e. neither under demand nor excessive demand of the cognitive processing possibilities, is the object of the cognitive ergonomics.

As quite in general for the whole use of ergonomic knowledge the following two approaches are also to be distinguished with regard to cognitive questions:

- The *retrospective measurement* of the goodness of an existing design solution concerning cognitive load, to be equated with the so-called corrective ergonomics and
- The *proactive design* of a system with the aim to minimize the cognitive load which corresponds to the so-called prospective ergonomics.

Presently in the area of the cognitive ergonomics still the retrospective approach controls the design process. After a system has been designed and created cognitive charges are measured (e.g., by questionnaires, user tests, simulations). This as *prototyping* known approach permits to clear cognition psychological questions only with existence of a system or at least a prototype (Balzert 1988). A reason for the dominance of the retrospective approach is the fact that cognitive loads are hardly predictable in a not closer specified technical system. However, on the other side those *post factum* investigations are always cost-intensive and are time consuming when the design solution must be reworked, because repeated going through of the design cycles forces them. Accordingly seldom technical systems are reshaped, even if it could be shown that they have led to events or accidents (Sträter and Van Damme 2004).

For this reason proactive creation methods were developed increasingly to the more efficient design concerning development time and development expenses (Hollnagel 2003). They are based on cybernetic, system-ergonomic approach which investigate the flow of information between person and machine concerning the in principle possible effort of code conversion (and with it of the cognitive load in principle to be expected).

Effort of code conversion can be distinguished in principle at two interaction levels (Balzert 1988):

- The relationship of the mental model of the user and the actual working principle of a technical system at the *representation level* or
- The relationship of the user desired and from the technical system demanded operating processes at the *dialogue level*.

This can be illustrated in the problem triangle shown in Fig. 8 which is put up between the poles system designer, actual realization and user to themselves. The system designer converts his internal imagination of the system and his possible use in a construction. The realization of it, which is not produced generally by the designer itself, cannot agree under circumstances by inserting other actors with the planned construction completely. The internal representation of the user of the arrangement can deviate from the realization as well as from the image of the designer. In the described discrepancies an essential source lies for misinterpretations and erroneousness operation.

Different cognitive control modes arise from this general connection that describes specific cognitive demands for the human operator (Hollnagel 1998). According to the stress and strain concept these cognitive demands lead to cognitive

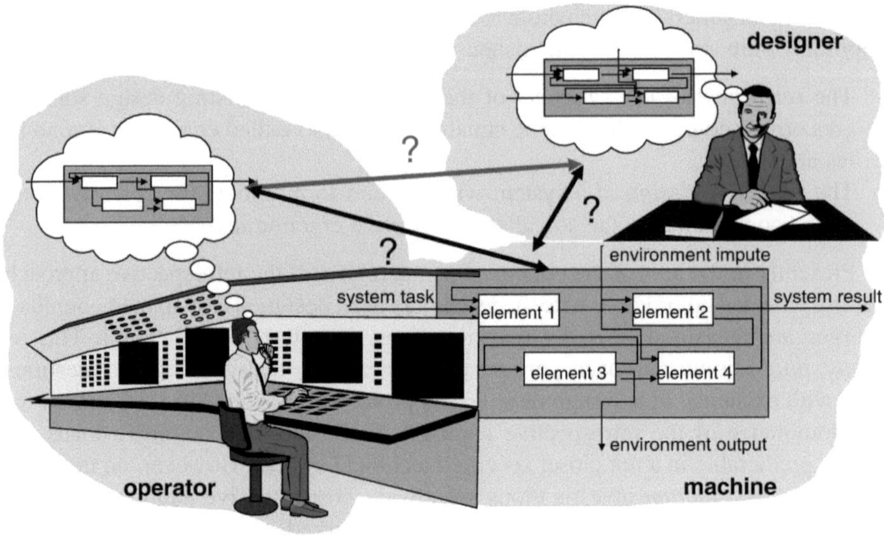

Fig. 8 Problem triangle between designer, realization of the system, and users

load. In a broadly invested research project to communication behavior in groups could be shown that these control modes and the cognitive loads following from them are also valid analogously for the human person's communication (GIHRE 2004; GIHRE stands for "Group interaction in High Risk Environments").

By the consideration of ergonomic design rules it is possible to give concrete advices how these demands can be fulfilled. These design rules are derived from system-ergonomic results of the research. They can be applied quite in general for the human operation design of machines.

For the ergonomic design of the task which arise from the system order and from the especially elective design of the system and the system components (e.g., of the software), the following basic rules which are formulated here as questions must be considered:

(a) *Function*: "What intends the operator and to what extent does the technical system support him?"
(b) *Feedback*: "Can the operator recognize whether has he caused something and what was the result of his action?"
(c) *Compatibility*: "How big is of the effort of code conversation for the operator between different canals of information?"

Figure 9 gives an overview, how the system-ergonomic design maxim "function", "feedback" and "compatibility" are further subdivided. Therefore, the function can be disassembled in the real task contents and the task design. How both names already imply, the contents arise from the logic basic of the task, while the task design is carried out by the system designer.

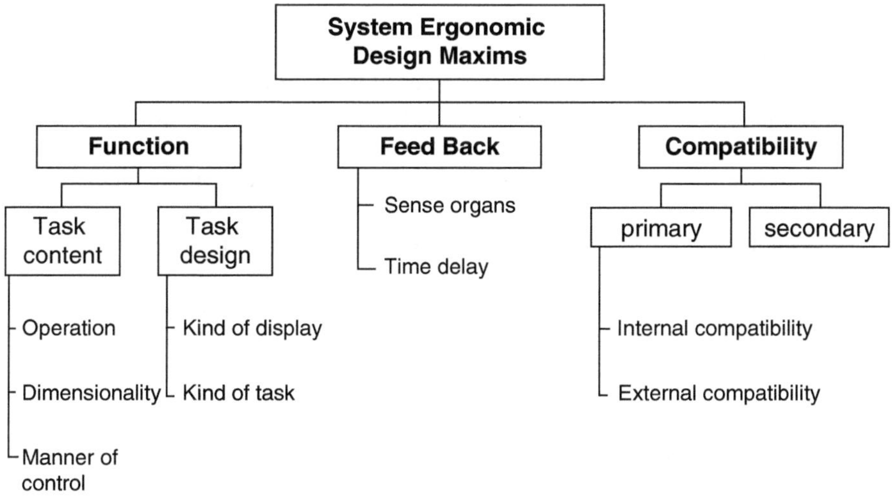

Fig. 9 Overview of the system-ergonomic design maxim function, feedback and compatibility

5.1 Function

The *function* refers on the one hand to the real *task content* that is fixed and describes substantially the change of information demanded by the human operator, and, on the other hand, on the *task design* that can be influenced within the scope of the respective technical possibilities by the system designer. The mental load of the operator is influenced by the complexity of the task, as well as by such aspects of the task which are determined by the provided technical system. Therefore the task of the system designer is to reduce the design-conditioned operating difficulty as much as possible by a favorable design, so that finally only the immanent of the task difficulty is left.

5.1.1 Task Content

Operation: Every task can be described by its temporal and spatial order of the necessary activities. For example, during the approach for a landing the aerodynamic lift decreased by the increasingly diminished speed must be compensated by moving out different landing airbrakes. Therefore, it is the task of the pilot to operate these landing airbrakes in a temporal sequence depending on the ongoing speed. Simultaneously he has to compensate different spatial air streams by separately operating the landing airbrakes of the right and left side of the airplane.

Thereby *temporal aspects* can be categorized by the differentiation in *simultaneous* and *sequential* operation. If the order of the necessary working steps is time wise given, one speaks of a sequential operation. It calls the fact that from objective point of view (not caused by the realized software!) certain processes in this and only this order are to be treaded (in the present example: Reaming of an undercut). If there is no essentially given temporal order of the working steps, a simultaneous process is given. "Simultaneous" means here an arrangement of equal standing of different choice possibilities, on this occasion, i.e. that different tasks stand in a queue at the same time and it is put in the pleasure of the operator in which order he accomplishes the tasks.

As soon as the user has the choice of several possibilities, a simultaneous operation is demanded by him. Altogether three different types, namely simultaneous operation of *compelling*, *varying* and *divergent kind* can be divided outgoing by these decisive possibilities (Rassl 2004). With the compelling kind several working steps stand in a queue at the same time for the decision. Each must be carried out to the fulfillment of the task. A typical pattern for a simultaneous operation of compelling kind is complete filling up a personal record. In this case the task consists in giving all data. Whether, nevertheless, first the given name or surname is put down on the record, is irrelevant for the result. Also with the varying kind different working steps stand in a queue at the same time for the decision. However, not each must be worked on because every possible step leads to the aim. To the fulfillment of the task it is enough to explain only one working step. An example is the choice of different route proposals in a navigation system (highway, yes/no, toll streets avoid, the quickest/ shortest distance etc.) which lead, however, in every case to the same destination. A simultaneous operation of divergent kind is given if the choice leads to different results. An example of it is the choice of destinations from the addressing memory of a navigation system. In Table 1 the three types of the simultaneous operation with a schematic flowchart in each case are summarized.

The temporal order of tasks is shown suitable-wise with a flowchart. In this flowchart by rhomb needs of information on the part of the machine (the machine cannot know what the operator wants) are characterized and by rectangles actions of the operator by which he transfers information on the machine. In the first step of an information ergonomic solution a so-called demand flowchart is to be created, which shows the necessary temporal order of the information transfer to the machine from the point of the operator without consideration of any technical realisation possibilities. Then from it is already to be derived, which tasks are simultaneous and which are sequential. In the case of simultaneous tasks it must be left to the operator which operating step he chooses first. The order of an essentially necessary sequence should be given on the other hand firmly in the software programme or by the arrangement of the control elements and the operator must have in suitable manner feedback about that on with which step the system just is.

The analysis of the operation is basic for every information ergonomic design. Now in the next analysis steps the respective ergonomic design must be carried out for each of the single actions which are characterized in the flowchart by rectangles.

Dimensionality: The difficulty of the *spatial order of* the task depends on the number of the dimensions on which the operator must have influence. A task is easy if only *one dimension* (e.g., the setting of a pointer on an analogous instrument) or *two dimensions* (e.g., the setting of a point on the screen of a computer must be

Table 1 Summary of three subspecies of the simultaneous operation. According to prevailing simultaneous choice possibilities this presents itself differently

Compelling kind	Several working steps stand in a queue for the simultaneous decision Every working step must be carried out Order of the processing is irrelevant for the result	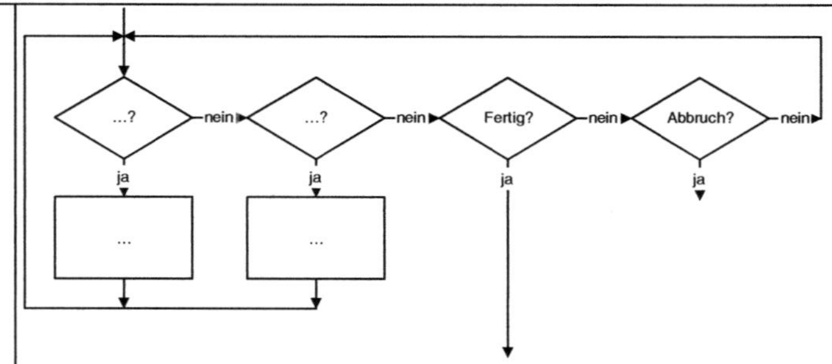
Varying kind	Several working steps stand in a queue for the simultaneous decision Only one working step must be carried out to the task fulfillment Every working step leads to the aim	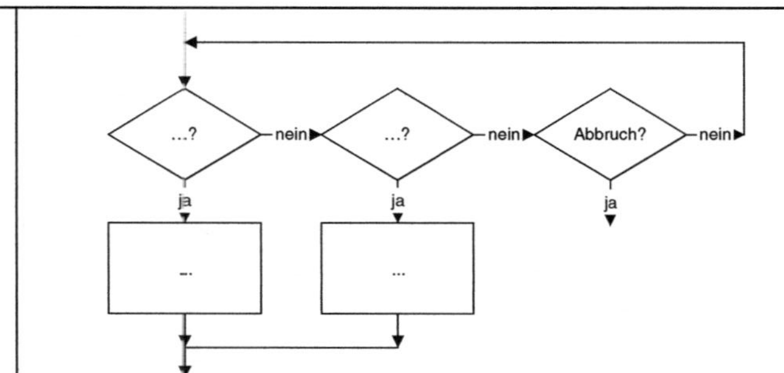

(continued)

Table 1 (continued)

Divergent kind	Several working steps stand in a queue for the simultaneous decision Not every working step must be carried out The single working steps lead to different results	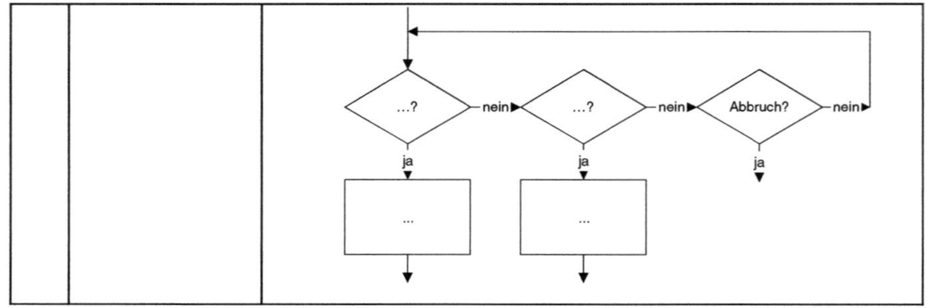

controlled by means of the mouse or the steering an automobile on two dimensional surface of the world) are to be controlled; in itself is a three-dimensional task also easy (e.g., piloting an airplane – the difficulty lies here in the control of the dynamic of the airplane), however, can become quite difficult by the necessary representation on a two-dimensional screen. In any case it is more difficult if it becomes necessary to control *four dimensions* (e.g. controlling a portal crane, as it is necessary to influence as the longitudinal as the transversal, the height movement as well as the orientation of the load around vertical axis) and especially difficultly if *six dimensions* must be influenced (e.g., positioning of a component for the purpose of the assembly or docking maneuver of a spaceship in the orbit).

Ergonomic improvements are reached if the number of the control elements is nearly immediately with the number of the influenced dimensions (e.g., the two-dimensional task of the motoring can become safer and more efficient if the control elements of a hand-switched car are substituted by an automatic, with only three control elements; the movement of a pointer on the two-dimensional screen surface by the two-dimensional movable mouse shows on the other hand almost an operating-ergonomic optimum. The control of an object in a quasi-three-dimensional representation on the screen by an operating element shows on the other hand an ergonomic challenge; the so-called space-mouse is an example of a good beginning in this direction).

Manner of control: In some cases the task is to be accomplished within a certain time window, what possibly time pressure originates, or within a certain local window (local window: the limited surface of the screen often makes a movement of the whole contents necessary. This leads to quite specific compatibility problems; see the following segment 5.1.2 and 5.3). A big time window characterises *static tasks* and a small time window *dynamic tasks*. Static tasks are characterised by instructions (nearly) independent of time with regard to the demanded end product (e.g., reading a value of a display; work on a lathe: only the final qualities of the work piece to be worked on are described by the drawing, not, nevertheless, the way, how this is produced). With time budget the relation required at the available time is called. It should not cross values more than 0.5. Then it is well possible to correct an observed mistake. Under this aspect it is to be rejected, by the way, ergonomically that certain operating setting are automatically put back after a certain time. In any case, better is a reset button deliberately to be operated by the operator. Dynamic tasks are characterized by the continuous service of a machine (e.g., driving a car on a winding street; directly operating a tool machine of the calculator console from).

5.1.2 Task Design

Kind of display: The difficulty of the task can be influenced furthermore by the way of the indication of task and result. In case of a technical indication the task and the result either can be indicated separately, or only its difference.

In the case of the separate indication one speaks of a *pursuit task*; it is recommended in observation situations which are earth-firmly installed, outside of vehicles (e.g., airplanes on the radar screen of the air traffic control or position of a marker in a CAD drawing). In this case the operator can win information about the movement or the change of the task and the result independently of each other. Therefore, he can make short-term predictions about the future movement of both and react consequently on time. Moreover, he thereby receives – with correct design of the compatibility (see Sect. 5.3) – a correct image of the reality on the display what makes easier the orientation to him.

In case of the indication of the difference of task and result one speaks of a *compensation task*. In technical systems this is often preferred because the display amplification can be freely chosen. Depending on the situation however, one exchanges this advantage by fundamental compatibility problems.

In connection with the operating surfaces of software one deals generally with the display of the pursuit task (by means of the mouse one can move to the single positions on the screen whereby their position does not depend on the mouse position). Because the screen window is limited by the size of the monitor, however, a movement of the representation is often necessary. Then one uses scrollbar. In this connection, there appears a compatibility problem which is very similar to the compatibility problem described below in vehicles: Is the position of the section or the position of the object behind the section moved by the scrollbar? The problem is solvable by the change to the pursuit task: while the cursor is brought on a free field (background) of the representation, the screen contents can be shifted with the pressed mouse key compatibly with direction.

The guidance of a car by the view on the natural surrounding is always a compensation task, because only the difference can be perceived between the own position and the desired position in the outside world. Therefore, the dimensions which deal with the outside world (e.g. representation of the correct course at a crossroad in a car by a navigation system; information of the run way in the case of the landing flight and the artificial horizon in the airplane) have to be shown in the form of a compensatory display from the view of information ergonomics. However, there arises in principle a compatibility problem (see Sect. 5.3), because with the steering movement the own vehicle is moved in relation to the fixed environment. This is presented to the driver of a vehicle, however, in such a way, as if the environment is moved in relation to the fixed vehicle. This discrepancy with the look at the natural environment ordinarily plays no role because the movement of own vehicle is natural for the driver of a vehicle. The example of the artificial horizon in the airplane and the difficulties, the pilots with it in the case of pure instrument flight have, however, shows the problem (see more in Fig. 10). The dilemma can be compensated by help of the contact-analogous indication in the HUD, because in this case the indication provides for the pilot of a vehicle the same information like the natural environment.

Kind of task: For the person operating the machine it is a basic difference whether he himself is integrated actively into the working process, or whether he has to fix only the basic settings of an automated process and then to observe the successful running. One speaks of *active* or *monitive* task of the person. Because

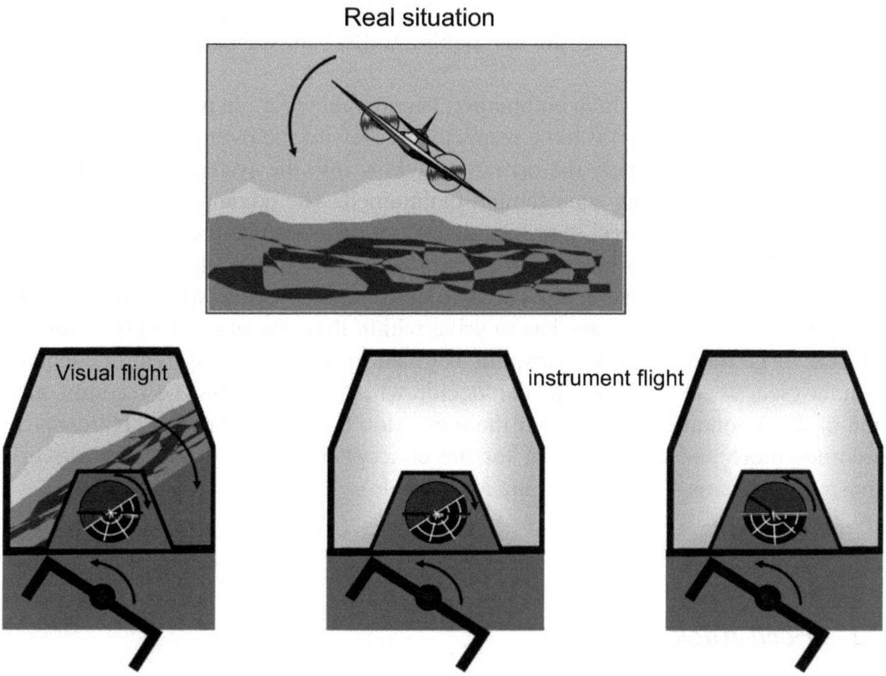

Fig. 10 Possible forms of presentation of the artificial horizon in the airplane. In the case of the instrument flight it comes to an incompatibility between the necessary movement of the steering gear and the reaction of the movable part of the artificial horizon (instrument flight on the *left*). If a movable airplane symbol is shown (instrument flight on the *right*), the movement direction of steering gear and the indication is compatible though, the movement of the airplane symbol, however, is incompatible to the movement of the outside view, in the case free sight conditions exist

these are almost the domain of the computer to regulate also a complicated process intelligently, today such automated processes win more and more in meaning.

Well known qualities of the person and those of the machine can be pulled up for the decision of the choice between automatic and manual control: Automation must be recommended in general if problems with human borders appear concerning the exactness, quickness and reliability. Nevertheless, this is correct only if exclusively such situations occur, for which the designer has made arrangements. Because for this, nevertheless, no guarantee can be given, a certain integration of the person is necessary in every system.

Monitive systems have disadvantages which arise from the risk of the monotony and lead therefore to a loss of the watchfulness of the operator. Moreover, the operator is by the constantly working automatic of the system relatively unskilled with the contact of the system elements and their impacts. In particular arises the disadvantage that the operator gets difficulties with the system control on account of his insufficient familiarity with the system in the case of a possible failure of the automatic. It arise the problem of the so-called situation awareness. In some cases because of the complexity of the situation and the impossibility to take up technical signals and to process (e.g. in the public traffic) an automation is not at all completely possibly.

With the systems in which automation is realizable (e.g., in most power stations and process observations) is to be respected in view of the overall reliability of the system to integrate actively the operator anyhow into the system because thus his attention and his training state remain maintained. The advantages of a monitive system for a reliable system in preplanned situations could be preserved in connection with this demand if only the local and temporal borders are determined by a machine within which the operator has to hold the dimensions to be influenced of the system. Then the operator has to drive within these borders the system always himself. If he touches the borders, this is given for him by the system in adequate manner (see “feedback”) and he can decide whether he follows the recommendation of the automatic or takes over the regulation himself (today this “philosophy” becomes more and more applied for the concept of so-called advanced driver’s assistance systems – ADAS – in the automobile).

5.2 Feed Back

The feedback of the achieved state to the operator is one of the most important factors by which a coherent understanding of the system state can be provided. If the information about the state of the system can be provided by different senses organs, this redundancy is to be evaluated in general positively. The human situation capture (“situation awareness”) rises if the same information is perceived by more than two senses organs (indication of a danger by lighting up of a control lamp *and* an acoustic signal). Another especially important aspect is the time span which exists between the input information in control element and the reaction of the system on the output side. If this time exceeds more than 100–200 ms (response time of the information perception) this leads to the confusion and disorientation of the operator. If such a time delay is not avoidable for technical reasons, this must be indicated to the operator (e.g., in the most primitive manner by the known “hourglass symbol”). If the time delay amounts more than 2 s., so the process to be regulated appears to the operator like a open control system. Then he needs at least the immediate feedback about the operated system input (e.g., high lighten of the button). However, the exact feedback about the progress of the process is better in this case (e.g., Indication by a growing beam and information of the rest time to be expected after initiation of a computer program).

If the points demonstrated here are followed, a well formed feedback always permits the answer to the questions to the operator:

- “What have I done?”
- “In which state is the system?”

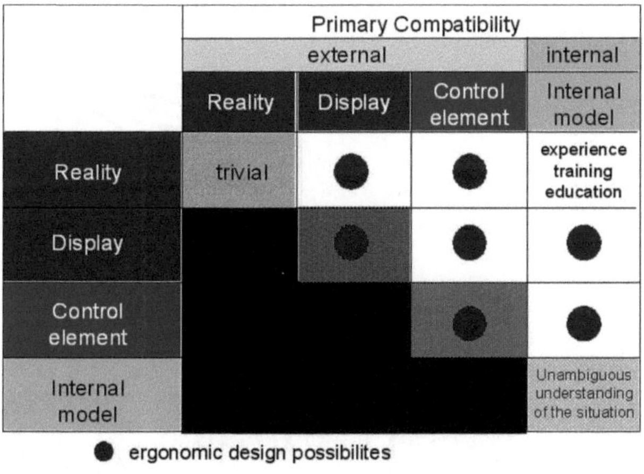

Fig. 11 Ergonomic design areas of the compatibility

5.3 Compatibility

Compatibility describes the easiness with which an operator is able to change the code system between different canals of information. Hereby is to be distinguished between primary and secondary compatibility.

Primary compatibility refers to the possible combinations of different areas of information like reality, displays, control elements and internal models of the operator (see Figure 11). Within the primary compatibility can be distinguished furthermore between external and internal compatibility: External compatibility refers to the interaction between person and machine concerning the outside world ("reality"; an example for this is the already discussed compatibility of steering gear and artificial horizon in the airplane; see Figure 10), while internal compatibility considers the interaction between the outside world and the suitable internal models of the person. As shown in Figure 11, an ergonomic design is possible only in certain areas. Thus, e.g., the compatibility of different information from the reality is trivial. On the other hand, compatibility can be reached between the reality and internal models only by experience, training and education. Compatibility between different internal models corresponds to an understanding free of contradiction of the situation. Every person has certain inconsistencies concerning the representations in different areas of his memory. The remaining areas in Figure 11 can be designed ergonomically in the sense that, for example, a movement forwards or to the right in the reality corresponds to a movement forward or to the right in the pointer or in the control element etc. This so-called spatial compatibility plays a prominent role in connection with the representation of the real outside world on a display (e.g., screen).

The internal compatibility is characterized by so-called stereotypes (see Fig. 12) that possibly only in the western society have this meaning (probably caused by the writing habit from the left to the right and from above down).

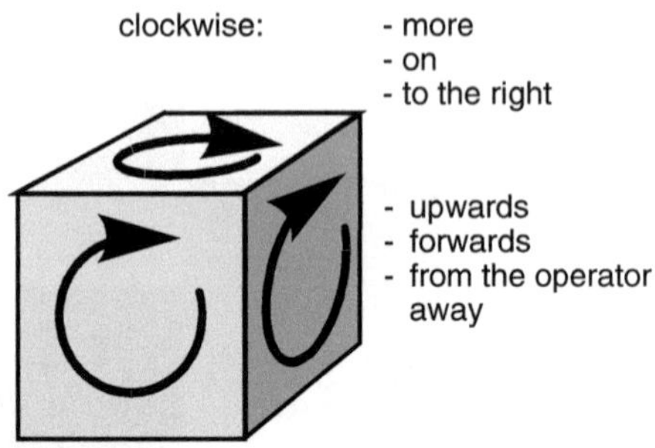

Fig. 12 Stereotypes

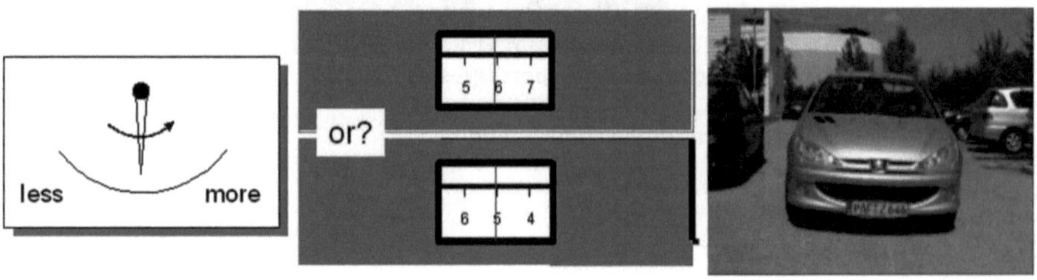

Fig. 13 Examples of secondary compatibility

Secondary compatibility means that an internal contradiction is avoided between partial aspects of the compatibility. Thus, for example, a hanging pointer is secondarily incompatible as there the movement from "on the left = a low" to "on the right = a much" (this in itself is compatible) is connected incompatibly with an anti-clockwise turn of the pointer. Also the case of "standing pointer – moved scale" is secondarily incompatible because here the movement direction of the scale is always incompatible to the arrangement of the order of rank of the digits. Another example is the arrangement of the indicator in the internal corner of the headlight of a car. In this case the position of the flashing light in the headlight case is incompatible to the position of the headlight in the vehicle. Figure 13 shows examples. Secondary incompatibility seems to be unimportant in the first sight because one understands in case of a rational consideration "nevertheless, easily how it is thought". However, in a critical situation, where it is important that information is perceived fast and reliably, just the secondary incompatibility can be occasion for a misinterpretation.

6 Conclusion

In the past especially ergonomics oriented in the physiology was more in the foreground by which unfavourable physical load could be recorded and reduced. In the next phase of the development the anthropometric ergonomics won importance by which the width of different body dimensions found consideration. Therefore unfavourable postures could be avoided; sight spheres and grasping spheres adapted to different sized individuals could be designed. Further on seat comfort could be improved. Excited by the rapid progress of the calculator technology today the information ergonomics comes in the focus of the consideration. It is a challenge of first rank to design the interaction of the intelligent person with intelligent machines. The progress of psychological research brings it with itself that the human information processing is understood better and better and by this can be modelled itself. Therefore, to a certain extent it is possible to precalculate the human behaviour in certain situations and to base on it the design of the interaction man and machine. Within the scope of the system ergonomics one has started very early developing rules for the creation of this interaction. These have of course also validity in connection with systems, which are controlled in excessive manner by the computer. Nevertheless, a change of gravity arises by the new possibilities above all also sped up by the sophisticated computer technology to visualise information. Thus information ergonomics wins importance as it is especially concerned with this topic. Some basic rules were shown in the present chapter which form the fundament for a information oriented design of the interaction man and machine.

References

Balzert, H. (Ed.). (1988). *Einführung in die Software-Ergonomie*. Berlin: de Gruyter.

Bubb, H. (1987): Entropie, Energie, Ergonomie; eine arbeitswissenschaftliche Betrachtung. München: Minerva.

Bubb, H. (2003a). Systemergonomie. [WWW-Dokument] Verfügbar unter: http://www.lfe.mw.tu-muenchen.de/index3.htm. [23.03.2004].

Bubb, H. (2003b). Ergonomie und Arbeitswissenschaft. [WWW-Dokument] Verfügbar unter: http://www.lfe.mw.tu-muenchen.de/index3.htm. [16.01.2004].

Bubb, H. (2006). A consideration of the nature of work and the consequences for the human-oriented design of production and products. *Applied Ergonomics, 37*(4), 401–408.

Bullinger, H.-J. (1990). Verteilte, offene Informationssysteme in der betrieblichen Anwendung: IAO-Forum, 25. January 1990 in Stuttgart. In H.-J. Bullinger (Hrsg.), *IPA-IAO Forschung und Praxis*: *T15* (S. 138–153). Berlin u.a.: Springer.

Bullinger, H.-J., Fähnrich, K.-P., & Ziegler, J. (1987). Software-Ergonomie: Stand und Entwicklungstendenzen. In W. Schönpflug & M. Wittstock (Eds.), *Software-Ergonomie*. Stuttgart: Teubner.

Crossmann, E. R. F. W., & Szostak, H. (1969). Man–machine-models for car-steering. *Fourth annual NASA-University conference on manual control National Aeronautics and Space Administration*, NASA Center for AeroSpace Information (CASI), Washington, DC.

DIN V 19 250. (1989). *Deutsches Institut für Normung*: *Messen, Steuern, Regeln*: *Grundlegende Sicherheitsbetrachtungen für MSR-Schutzeinrichtungen, Vornorm*. Berlin: Beuth-Verlag.

Döring, B. (1982). System ergonomics as a basic approach to man–machine-systems design. In H. Schmidtke (Ed.), *Ergonomics data for equipment design*. New York: Plenum.

Döring, B. (1986). Systemergonomie bei komplexen Arbeitssystemen. In R. Hackstein, F.-J. Heeg, & F. V. Below (Eds.), *Arbeitsorganisation und neue Technologien* (pp. 399–434). Berlin: Springer.

Gengenbach, R. (1997). Fahrerverhalten im Pkw mit Head-Up-Display. Gewöhnung und visuelle Aufmerksamkeit. VDI Fortschrittsbereichte Reihe 12: Verkehrstechnik/Fahrzeugtechnik. Nr. 330.

GIHRE. (2004). *Group interaction in high risk environments*. Aldershot: Ashgate.

Herczeg, M. (1994). *Software-Ergonomie: Grundlagen der Mensch-Computer-Kommunikation*. Oldenbourg: Addison-Wesley.

Hilf, H. (1976). *Einführung in die Arbeitswissenschaft* (2nd ed.). Berlin/New York: De Gruyter.

Hollnagel, E. (1998). *Cognitive reliability and error analysis method – CREAM*. New York/ Amsterdam: Elsevier. ISBN ISBN 0-08-042848-7.

Hollnagel, E. (Ed.). (2003). *Handbook of cognitive task. design*. New Jersey: Erlbaum.

Hoyos, G. C. (1974). Kompatibilität. In H. Schmidtke (Ed.), *Ergonomie 2*. München/Wien: Hanser Verlag.

Jürgensohn, T. (2000). Bedienermodellierung. In K.-P. Timpe, T. Jürgensohn, & H. Kolrep (Eds.), *Mensch-Maschine-Systemtechnik – Konzepte, Modellierung, Gestaltung, Evaluation*. Düsseldorf: Symposion Publishing.

Kahlisch, T. (1998). Software-ergonomische Aspekte der Studierumgebung blinder Menschen. Hamburg: Kovac 1998: Ill. – Schriftenreihe Forschungsergebnisse zur Informatik; 37 Zugl.: Dresden: Technische Universität (Dissertation).

Kraiss, K.-F. (1989). Systemtechnik und Ergonomie. Stellung der Ergonomie in der Systemkonzeption. In H. Schmidtke (Ed.), *Handbuch der Ergonomie*. Koblenz: Bundesamt für Wehrtechnik und Beschaffung.

Luczak, H. (1998). *Arbeitswissenschaft* (2nd ed.). Berlin: Springer.

Marchionini, G., & Shneiderman, B. (1988). Finding facts vs. browsing knowledge in hypertext systems. *IEEE Computer, 21*, 70–90.

Mayo, E. (1950). *Probleme industrieller Arbeitsbedingungen*. Frankfurt/Main: Verlag der Frankfurter Hefte.

McRuer, D. T., & Krendel, E. S. (1974). Mathematic Models of Human Pilot Behaviour. AGAR Dograph No. 188, Advisory Group for Aerospace Research and Development, Neuilly sur Seine.

Mutschler, H. (1982). Ergonomische Aspekte der Spracheingabe im wehrtechnischen Bereich. In K. Nixdorff (Hrsg.), *Anwendung der Akustik in der Wehrtechnik*. Tagung in Meppen, 28–30 Sept 1982.

Pöppel, E. (2000). *Grenzen des Bewußtseins – Wie kommen wir zur Zeit und wie entsteht Wirklichkeit?* Frankfurt a.M.: Insel Taschenbuch.

Rasmussen, J. (1987). The definition of human error and a taxonomy for technical system design. In J. Rasmussen et al. (Eds.), *New technology and human error*. New York: Wiley.

Rassl, R. (2004). Ablenkungswirkung tertiärer Aufgaben im Pkw – Systemergonomische Analyse und Prognose. Dissertation. München: Technische Universität München.

Schmidtke, H. (1989). Gestaltungskonzeption. Ziele der Gestaltung von Mensch-Maschine-Systemen (S. 2). In H. Schmidtke (Hrsg.), *Handbuch der Ergonomie*. Koblenz: Bundesamt für Wehrtechnik und Beschaffung.

Schmidtke, H. (1993). *Ergonomie* (3rd ed.). München/Wien: Hanser Verlag.

Schneeweiß, H. (1967). *Entscheidungskriterien bei Risiko*. Berlin: Heidelberg/New York.

Simon, S., & Springer, J. (1997). Kommunikationsergonomie – Organisatorische und technische Implikationen zur Bewältigung von Informationsflut und -armut. In *Jahresdokumentation der Gesellschaft für Arbeitswissenschaft* 1997 (S. 149–152). Köln: Otto Schmidt Verlag, (SV 2796).

Stein, M. (2008). Informationsergonomie – Ergonomische Analyse, Bewertung und Gestaltung von Informationssystemen. Habilitationschrift.

Stein, M., & Müller, B. H. (2003a). Informationsergonomie – Entwicklung einer Checkliste zur Bewertung der Nutzerfreundlichkeit von Websites (S. 41–44). In H.-G. Giesa, K.-P. Timpe, U. Winterfeld (Hrsg.), *Psychologie der Arbeitssicherheit und Gesundheit,* 12. Workshop 2003. Heidelberg: Asanger Verlag.

Stein, M., & Müller, B. H. (2003b): Maintenance of the motivation to access an internet-based information system. In: *Zentralblatt für Arbeitsmedizin, Arbeitsschutz und Ergonomie*, Heft 5, Bd. 53, S. 214. Heidelberg: Dr. Curt Haefner-Verlag.

Straeter, O. (1997). *Beurteilung der menschlichen Zuverlässigkeit auf der Basis von Betriebserfahrung. GRS-138*. Köln: GRS. ISBN 3-923875-95-9.

Sträter, O., & Van Damme, D. (2004). Retrospective analysis and prospective integration of human factor into safety-management. In K.-M. Goeters (Ed.), *Aviation psychology: Practice and research*. Aldershot: Ashgate.

Turing, A. M. (1936). On computable numbers, with an application to the Entscheidungsproblem. *Proceedings of the London Mathematical Society, 42*, 230–265.

Volpert, W. (1993). *Arbeitsinformatik – der Kooperationsbereich von Informatik und Arbeitswissenschaft. Zeitschrift für Arbeitswissenschaft (2)*. Köln: Verlag Dr. Otto Schmidt KG.

Wahlster, W. (2002). Zukunft der Sprachtechnologie. Vortrag auf Veranstaltung Bedien-und Anzeigekonzepte im Automobil des Management Circle, Bad Homburg, 04.02–05.02.2002.

Wunsch, G. (1985). *Geschichte der Systemtheorie, Dynamische Systeme und Prozesse*. München/ Wien: Oldenbourg.

Zeidler, A., & Zellner, R. (1994). *Softwareergonomie, Techniken der Dialoggestaltung*. München/ Wien: Oldenbourg-Verlag.

Zülch, G. (2001a). Kommunikationsergonomie. [WWW-Dokument] Verfügbar unter: http:// www.ifab.uni-karlsruhe.de. [19.01.2004].

Zülch, G. (2001b). Forschungsschwerpunkte/Kommunikationsergonomie. [WWW-Dokument] Verfügbar unter: http://www.rz.uni-karlsruhe.de/~ibk/FEB/VIII_2_d.html. [16.03.2004].

Zülch, G.; Stowasser, S; Fischer, A. E. (1999): Internet und World Wide Web - zukünftige Aufgaben der Kommunikationsergonomie. Ergonomische Aspekte der Software-Gestaltung, Teil 9. In: Ergo-Med. Heidelberg 23, 1999 2, 6–100.

Zwahlen, H. T. (1988). Safety aspects of CRT touch panel controls. In A. G. Gale (Ed.), *Automobiles, vision in vehicles II*. Amsterdam: Elsevier.

Human Aspects of Information Ergonomics

Michael Herczeg and Michael Stein

1 Introduction

The need for information within a real-time operational environment, e.g. a transportation system, is related to predefined tasks and emerging events as well as the resulting activities and communication processes. Human operators have to perform such tasks and react to such events taking place in their environment. They have to interact and communicate with other humans and machines to fulfill their duties.

Especially in the field of transportation systems, information required by the operator for process control and regulating activities is to an increasing degree no longer perceived directly from the environment (outside view) but collected via sensor systems, processed, enriched with additional information, and communicated by media (Herczeg 2000, 2008, 2010; Stein 2008). Examples of this are head-up displays in which the real view is overlaid with symbols (scene-linked as required), forward looking infrared (FLIR), or also enhanced and synthetic vision systems (e.g. tunnel in the sky) consisting only of a synthetic image including additional information. As compared to the "mere" outside view in combination with "conventional" displays, this shall enable the operator to gain more adequate situational awareness (SA) of the environmental and system status. The workload shall also be optimized and brought to an appropriate level (Nordwall 1995). The ultimate goals are the safe operation of the transportation systems as well as an enhancement of performance.

The media-based communication of information as described is accompanied by an increasing degree of automation (cf. modern aircraft: glass cockpit). In the sense

M. Herczeg (✉)
Institute for Multimedia and Interactive Systems, University of Luebeck, Luebeck, Germany
e-mail: herczeg@imis.uni-luebeck.de

M. Stein
German Air Force Institute of Aviation Medicine, Manching, Germany

M. Stein and P. Sandl (eds.), *Information Ergonomics*,
DOI 10.1007/978-3-642-25841-1_3, © Springer-Verlag Berlin Heidelberg 2012

of cognitive automation (cf. Flemisch 2001) information systems on an information presentation level are intended to enable (partial) congruity between the real process and the cognitive model of the process which is sufficient for the operator to perform the process control tasks successfully (Herczeg 2006b). This generally involves a system complexity which is partially presented in such a way that the cognitive resources of the operator are sufficient. If the system complexity exceeds the cognitive capacity of the operator he will usually respond with cognitive reduction strategies. This means that the operator will consciously or subconsciously choose parts of the complex system (or of a complex information system) which are of lower priority to him compared to other parts and he will eliminate these from his "cognitive model and plan". This can initially reduce the working memory load. However, if such action plans are needed, for instance in cases of emergency or also during phases of high cognitive work load, the operator usually cannot restore and apply these, possibly causing a loss of situation awareness or also a breakdown of communication and action regulation.

The increasing media-based communication of system and control relevant information as described above, combined with a large degree of automation and system complexity, create high demands especially of media-based communication of information. This complex communication patterns require sophisticated information ergonomics, in particular and illustrates the significance of information ergonomics, in particular the importance of modeling cognitive processes which control access to an information system.

This chapter contains an overview of how human operators accomplish these challenges physically and mentally. We will present and discuss some basic cognitive models of human activities and their limits as well as psychological constructs such as motivation, emotion, effort regulation, workload and fatigue.

2 Cognitive Models for Activities and Communication

Activity and communication processes have been described in several abstract models by many authors before. Some of these well-known models, which have been successfully used and proved in many application domains, will be described in the following sections. Particular emphasis will be placed on information perception, memorization, and processing as well as the generation of information within these processes.

2.1 Skills, Rules and Knowledge: Signals, Signs and Symbols

In accordance with Rasmussen (1983, 1984) the problem-solving activities of human operators mainly perform on three different levels of physical and mental capabilities. These levels represent different problem-solving strategies (cf. Fig. 1):

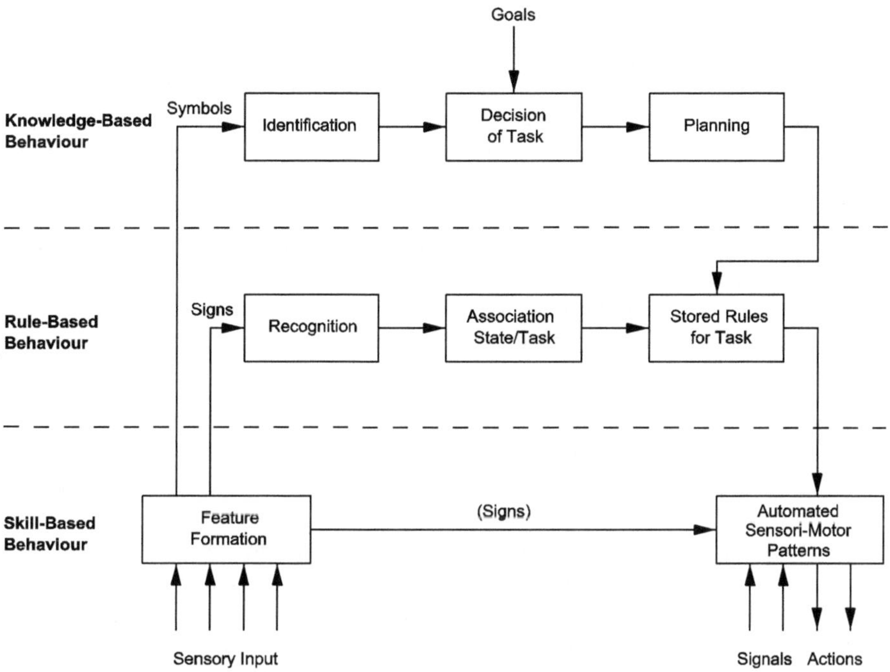

Fig. 1 Levels of problem-solving behaviour (cf. Rasmussen 1983)

1. *Skill-Based Behaviour*: perceiving events in the environment through signals and reacting with mostly automated sensomotorical procedures;
2. *Rule-Based Behaviour*: perceiving state changes recognized through signs and reacting with learnt rules;
3. *Knowledge-Based Behaviour*: perceiving more complex problems through symbols, solving these problems in a goal-oriented way and planning activities to react to the environment.

Rasmussen relates these three forms of behaviour to three forms or qualities of information perceived:

1. *Signals*: perceived physical patterns (data); e.g. the brake light of a car to inform the following driver about a brake maneuver;
2. *Signs*: processed signals relating to system states (information); e.g. the fuel indicator with a low-level warning light for the fuel tank of a car;
3. *Symbols*: processed signs relating to meaningful subjects, objects and situations in the environment (knowledge); e.g. the traffic information about a congestion some distance down the highway.

These three information qualities can be differentiated by specific characteristics. *Signals* are simple perceptions without meaning. They can trigger automated behaviour, usually stored in the subconscious domains of the human memory.

This automated behaviour (human *skills*) is a kind of compiled information supporting pattern recognition and fast reactions. *Signs* are selectors and triggers for predefined *rules*. Rules are highly structured knowledge in the form of condition-action-pairs where the signs are the condition parameters (keys) of the rules. *Symbols* are semantically rich information, often culturally loaded with various associations to people, objects and situations in the world perceived. They can be viewed as the building blocks of human *knowledge* (so-called *chunks*).

Process control systems must be designed in such a way that the information presented corresponds to the cognitive level of problem solving. This means that signals need to be safely discriminated initiating patterns for the precompiled, i.e. trained and memorized subconscious procedures. *Signals* have to meet a specific information quality, i.e. they:

- Have to be presented with a signal to noise ratio that prevents both alpha-errors (not recognizing the signal) as well as beta-errors (erroneously perceiving a non-existing signal) (Wickens and Hollands 2000); e.g. the traffic light should not be perceived as being lit due to reflected sunlight;
- Must have a sufficient perceptual distance between one another because they will not be reflected and processed by higher-order knowledge for disambiguation; e.g. the brake light should not be confused with the backlight of a car;
- Should have sensory properties to support neural processing without preceding learning effort, rather than just being arbitrary (Ware 2000, p. 10 ff., Ware calls them "symbols", but refers to low level neuronal processing);
- Have to be presented within a time frame that allows for safe and timely reaction, e.g. the ABS feedback will be haptically fed into the braking motion of the driver's foot instead of providing a light or a warning tone which might be recognized and processed too late into an activity.

While signals support the lowest level of fast and skilled reaction in the sense of a pattern-matching construct, signs have to be interpreted in the context of system states. Certain system states will be preconditions for rules to be triggered. Signs play the role of additional triggering parameters for an operator to change a system state by some activity. Therefore, *signs* should have the following qualities, i.e. they:

- Should be perceived as system states, not just as perceived patterns; e.g. the oil pressure warning lamp provides the visual indication of an oil can;
- Have to represent certain system properties without ambiguity; the low fuel light should be clearly distinguishable from the low oil indicator;
- They have to be stable to allow a safe *rule matching* process (Newell and Simon 1972), i.e. enough time to check a potentially long list of rules to find the matching ones; the rule-matching process can be accelerated by preordering or compiling (i.e. changing their representation for efficient matching) rules according to their triggering signs; therefore, signs have an important discriminator function for actions to be taken, which means that they play an important role in structuring state information in the operator's mental models, e.g. there has to be a stable hysteresis for a low fuel light instead of a flickering warning light changing its state depending on the current driving dynamics.

Whenever there are no matching skill patterns or rules, signs can be processed further loaded with further meaning to be perceived as symbols within a framework of knowledge representation. *Symbols* can therefore be:

- Larger information structures characterizing a complex situation; e.g. a colored weather radar display showing a larger area in front of the airplane;
- Information bound to scenarios of dynamic problem solving; e.g. a TCAS display presenting not only the collision danger, but the direction to be taken as well;
- Signs related to objects, their relations and meaning, i.e. semiotic structures (Nöth 2000); a fuel tank display in an airplane shows all tanks positioned according the physical structure (layout) of the plane;
- Ambiguous information that has to be discriminated and used within complex contexts, e.g. radar echoes associated with transponder information to filter out irrelevant static objects for an ATC screen.

As a result, symbols cannot just be displayed on a screen. They must emerge as complex perceptual structures related to situations, behaviour, relations, activities, social relationships and even cultural contexts.

2.2 Layers of Functional Abstraction and Decomposition

Complex systems and processes with hundreds or even tens of thousands of parameters and multiple interrelations between these parameters cannot be conceived as a whole and cannot be understood on one level of abstraction. Rasmussen (1983, 1985a) proposed an *abstraction hierarchy* to reflect on several layers of abstraction (cf. Fig. 2) to understand and operate a complex process control system depending on the current context. The abstraction layers defined represent different ways of perceiving and understanding the functions of complex systems:

1. *Functional Purpose*: This level is related to the ultimate reason why the system has been built and put into operation. For a vehicle it will be its main function of transporting people or payload from one place to another. Typical information related to this level will be, for example, efficiency, time to destination or cost of delivery.
2. *Abstract Functions*: Looking one level deeper into the functional structure we will find more or less complex topologies of flow. In transport systems we will find for example the complex structure of routes leading from a starting point to a destination. The information needed on this level might be the topological structure and the capacity or the current load of different routes.
3. *Generalized Functions*: From this point of view we will find concepts like one-way or multilane routes, which are generalizable concepts for routes that can be used by one or more vehicles in the same or opposite directions in parallel or sequentially according to certain traffic modes.

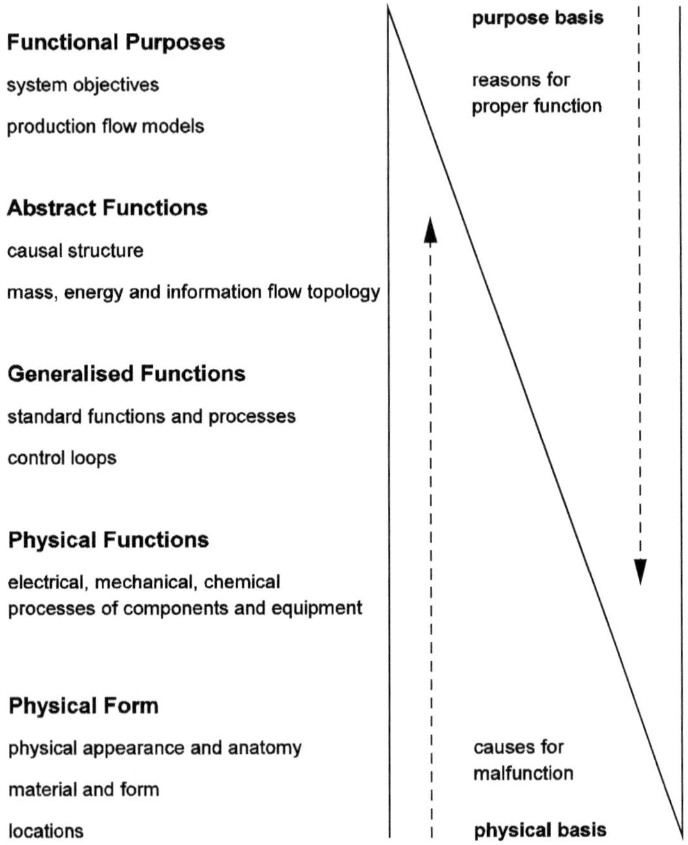

Fig. 2 Levels of abstraction for process control (cf. Rasmussen 1983)

4. *Physical Functions*: This level tries to map the general functions to a specific physical modality. A route for a ship will be a surface waterway, while a route for an airplane will be a 3D trajectory in the air. While it may make no difference to calculate time to destination on the level of generalized functions, the physical medium must be taken into account for calculations on this physical level.
5. *Physical Form*: The lowest level of abstraction deals with the final physicality of a process element. This might be, for example, the form, size and physical state of a road or the weather conditions of a selected flight level. Information on the physical level will be naturally perceived by multiple human senses. To give an operator access to this level he or she needs to be connected to the physical situation by multimodal perceptual channels, for instance visual, auditory or tactile senses (Herczeg 2010).

With respect to information ergonomics these abstraction layers can be used to provide a goal-oriented and problem-specific view of the process for different

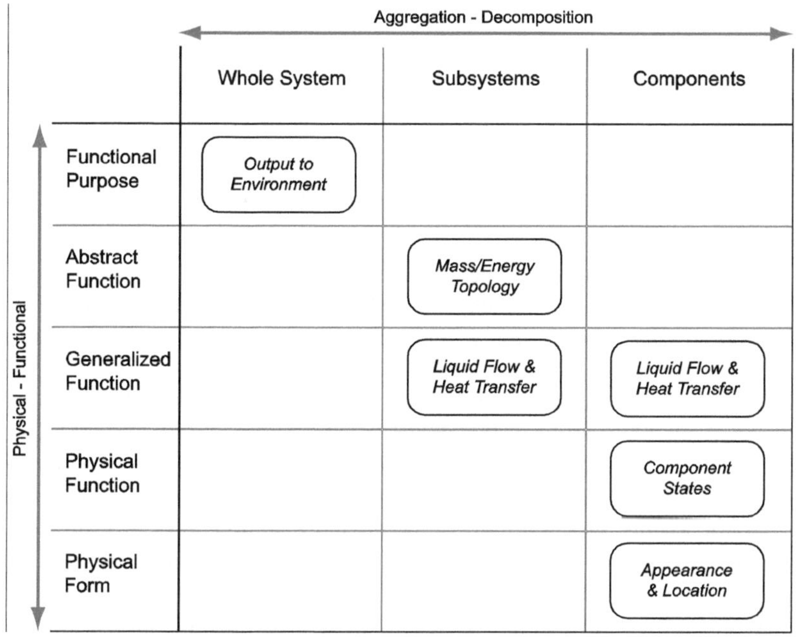

Fig. 3 Abstraction and decomposition (cf. Vicente 1999, p. 318)

operational purposes and contexts. For example, it will be necessary to view the physical situation if there is a need of action on this level. To check whether a railway track is physically usable after a blizzard, one needs to see the physical condition of the iron track. To decide whether a train may enter a station, one needs to check closed and reserved routes and their availability on the generalized functional level.

In addition, the abstraction hierarchy can be extended into a second dimension called the *decomposition hierarchy* (Rasmussen 1985a; Rasmussen et al. 1994). Decomposition can be understood as the multilevel hierarchy of a system starting at the overall system and leading down stepwise to its subsystems and final components (Fig. 3). It allows us to focus on a special part and its technical properties. For example, if there is a problem with the availability of an engine during flight, the pilot will need to focus on all engines, their power states and the fuel available, but not on the state of other subsystems in the airplane like the air-condition system or the landing gear.

The utility and usability of information based on abstraction and decomposition structures can be defined by a systematic reduction of complex information with respect to a system view (structural decomposition) as well as a process view (functional abstraction). The nodes in this two-dimensional grid must be associated with states, events and tasks to support the human operator in the recognition of situations as well as the planning and execution of activities. In most process control systems we will find examples of the usage of these abstraction and decomposition concepts. However, the information provided in a real system is rarely structured systematically according to these dimensions because of information missing, high development effort or other reasons.

2.3 Decision Ladder for Event-Based Activities

One of the main situations of operators in highly automated environments is the observation of the dynamic processes and the readiness for handling of events that emerge more or less unexpectedly. The process of perceiving events and reacting with appropriate actions can be described as a multi-staged cognitive processing pipeline which has been called the *decision ladder* by Rasmussen (1985b). We have already seen one perspective of this decision ladder when we discussed skills, rules and knowledge combined with the information qualities signals, signs and symbols. The decision ladder can be presented in greater detail during the cognitive process of perceiving events, processing information and planning and executing activities to bring the system from an abnormal state to back to normal operation (cf. Fig. 4).

With respect to information ergonomics, the main question is how signals can be enriched by further information and insight about their meaning until a system state can be derived from these perceptions. Having identified a system state the operator is usually able to recognize whether this state is an acceptable safe state or whether it is abnormal and needs to be regulated and changed into some goal state for normal operation. The main information enrichment steps in this perceptual process are:

1. *Activation*: detection of a need for information processing;
2. *Observation*: gathering of data and changing these data into perceived information;
3. *Identification*: derivation of a system state from this perceived information;
4. *Interpretation*: effects of the system state for operation and safety;
5. *Evaluation*: acceptability of the system state or safe procedures for its regulation.

The quality of a control system stems partially from its capabilities to present data to fulfill these steps in a timely and correct manner. The operator has to be supported in the perception and memorization process before any reasonable activities to handle the event can take place. Whenever this chain of informational perception will lead to an early recognition of actions that have to take place, the operator might apply shortcuts performing these activities (cf. Fig. 4). While going for these shortcuts the operator will leave out the complete bottom-up analysis and will try to react as fast and direct as possible. Whether these shortcuts will be adequate and lead into a safe goal state depends on the quality of information perceived and on the experience of the operator. Weak information and low competencies will lead to suboptimal or incorrect activities that might even aggravate the situation.

Information systems within transport systems must inform about system states and guide the regulation from abnormal to normal operation. Each stage of the

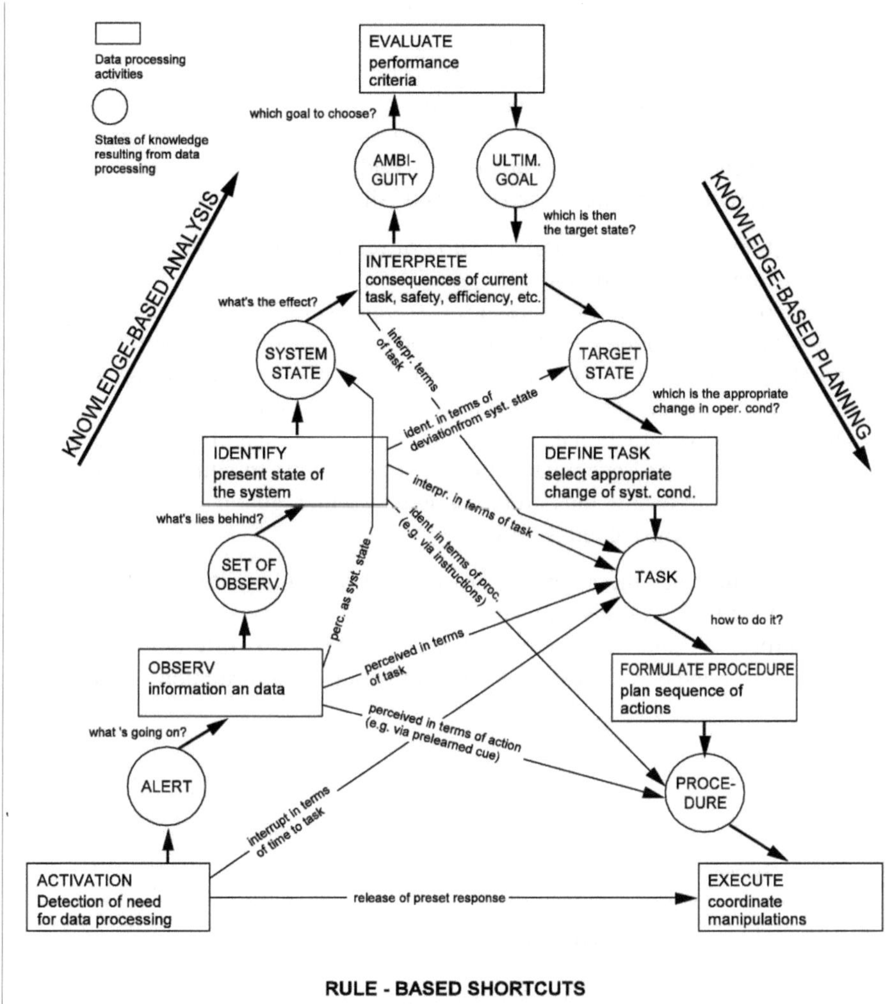

Fig. 4 Decision ladder for process control (cf. Rasmussen 1985b)

decision ladder requires a different information quality. Given information will be enriched and focused until the perception of system states and regulatory activities emerge.

2.4 *Situational Awareness and Decision Making*

While Rasmussen's decision ladder concentrates on a current and a goal state, there is no explicit consideration of the operator's anticipation of a future state. This aspect has been taken into account in the model of situation awareness (SA) by Endsley (1988).

One definition of situational awareness specified by Endsley (1988) and cited in numerous publications is: "*The perception of the elements in the environment within a volume of time and space, the comprehension of their meanings, and the projection of the status in the near future*" (Endsley 1988, p. 97). In this context Endsley differentiates between three levels of situational awareness:

- The first level comprises the *perception* of relevant cues from the environment (also media-based information).
- The second situational awareness level *(comprehension)* includes the recombination of the perceived cues to form a conclusive picture and the interpretation and storage of this picture with respect to their relevance regarding goals (the forming of a correct picture of the situation is based on perception).
- The third level of situational awareness describes the capability of *projecting* the "*conclusive picture*" generated in the first two levels to situations, events and dynamics in the future, thus initiating actions which guarantee and ensure high quality of performance and safety.

Figure 5 illustrates the cognitive processes determining situational awareness. Endsley (1995) clearly distinguishes between situational awareness, decision-making processes and performing actions. Endsley also differentiates between situational awareness as a result of the acquisition of information and receiving

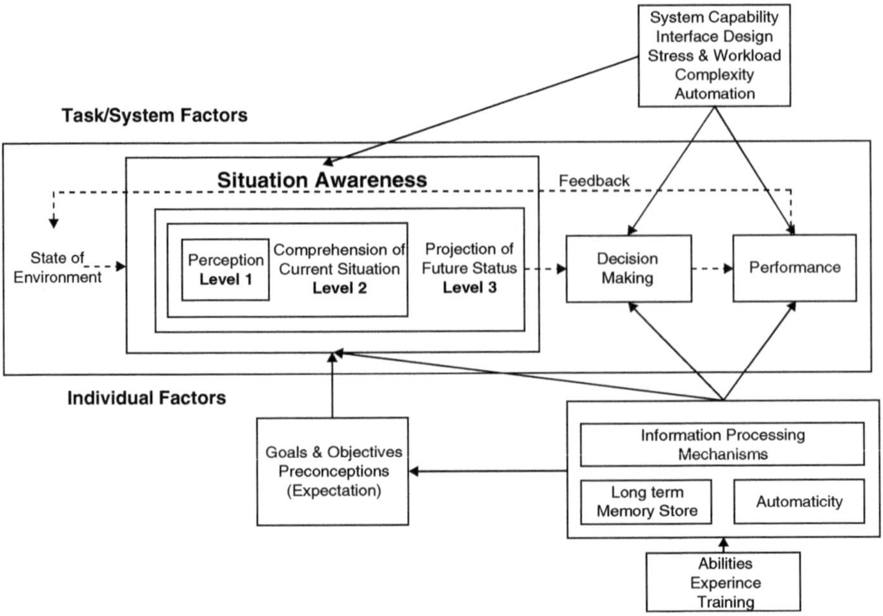

Fig. 5 Model of situation awareness (Endsley 1995)

information as a process she calls situation assessment. She further explains her situational awareness model, declaring that situational awareness is the main precursor for decision making; however, like decision making and performance of actions, situational awareness is affected by moderating variables, e.g. stress, workload and complexity. Waag and Bell (1997) disagree with this perspective and regard situational awareness in terms of a meta construct in which information processing, decision-making and execution of action are an integral part (cf. Seitz and Häcker 2008).

The limiting variables in the situational awareness process are the limited resources of attention (cf. data-limited and resource-limited processes; Norman and Bobrow 1975) and working memory. The experience of the operator plays a decisive role in dealing with these limitations. Experienced operators actively scan the environment by means of an elaborate cognitive model (in terms of top-down processes, i.e. goal-directed) for action-relevant information. The prioritization between different competitive goals in dependency on the situation seems to be significant. Such highly automated information processing saves the working memory resources on the one hand and contributes to developing more adequate situational awareness on the other.

There are various decision-making processes with regard to the use of information systems within the scope of transportation. The decision:

- Of whether to access an information system (Does the system and environmental state allow attention to be directed to the information system? Is access to the information system required? Is e.g. the "total" situational awareness lost?);
- To be made at what point in time the information system shall be accessed (some phases in a dynamic process can be more or less advantageous for access);
- To be made with regard to how long access to the information system shall last (The operator generally has an idea of how long he can divert his attention from main controls or the monitoring task. Generally he also has an idea of how long the search for information will take. Due to the fact that the system or environmental state changes during the use of the information system a change of goals may result during the search for information.);
- On which information is required (prioritization, if required);
- On the application or rejection of information. (Does the user make a decision on how to act based on the information received or does he make his decision based on other information; Stein 2008?).

A distinction can be made between internal and external situational awareness with regard to the use of information systems within the scope of transportation. External situational awareness means all three situational awareness levels with regard to the system and environment state, e.g. internal situational awareness refers to the use of information systems. Especially in modern aircraft (glass cockpit) the operator must perform less and less control and regulating activities but more and more monitoring tasks. Frequently he only intervenes in terms of regulating and controlling the system if certain critical constellations occur, requiring his action. In this regard information systems, e.g. electronic flight bags, are applied. These information systems generally consist of a complex setup of structures similar to hypertext (or hypermedia) as well as several information levels. With regard to use, inner situational awareness means that the pilot knows which part of the information system he is accessing, how to obtain certain information and how to return to the original state.

With regard to the development of the situational awareness construct Seitz and Häcker (2008) currently declare that – similar to the emotional intelligence construct – this has not been fully differentiated and completed. In addition, the authors criticize that there are two deviating concepts of the situational awareness construct and there is no standardization of situational awareness measures. Furthermore, Seitz and Häcker (2008) criticize that many investigations were planned and conducted without reviewing already existing literature. As a consequence, investigational hypotheses are generally not derived based on theory but are "generated" due to operational reasons. This does not contribute to a stringent further development and differentiation of the construct and the theory.

2.5 Action and Perception for Task-Based Activities

While we discussed event-based operations, operators will usually perform predefined or even scheduled tasks. In such situations operators refine main operating goals from a high-level task down to the level of physical activities (Norman 1986; Herczeg 1994, 2006a, c, 2009). High-level goals in the field of aviation are, for example, take-off, en-route autopilot flight, change of flight level or landing. Examples of high-level tasks in driving a car are starting the car, parking a car or overtaking another car. Starting with the high-level goal there will be a stepwise refinement of activities through the following levels (cf. Fig. 6 left side branch):

1. *Intentional Level*: high-level goal;
2. *Pragmatical Level*: predefined or developed procedures to reach the goal;
3. *Semantical Level*: state changes to be performed with respect to objects in the problem domain;
4. *Syntactical Level*: activities to be performed with the control system according to the operating rules;
5. *Lexical Level*: refinement of the syntactic rules into the multiple levels of super signs (aggregation of signs) and elementary signs (keys, buttons, icons, names, etc.);
6. *Sensorimotorical Level*: physical actions to execute a step.

After the activities have been performed on the physical level, the control system will receive the commands, process them, react and provide feedback to the

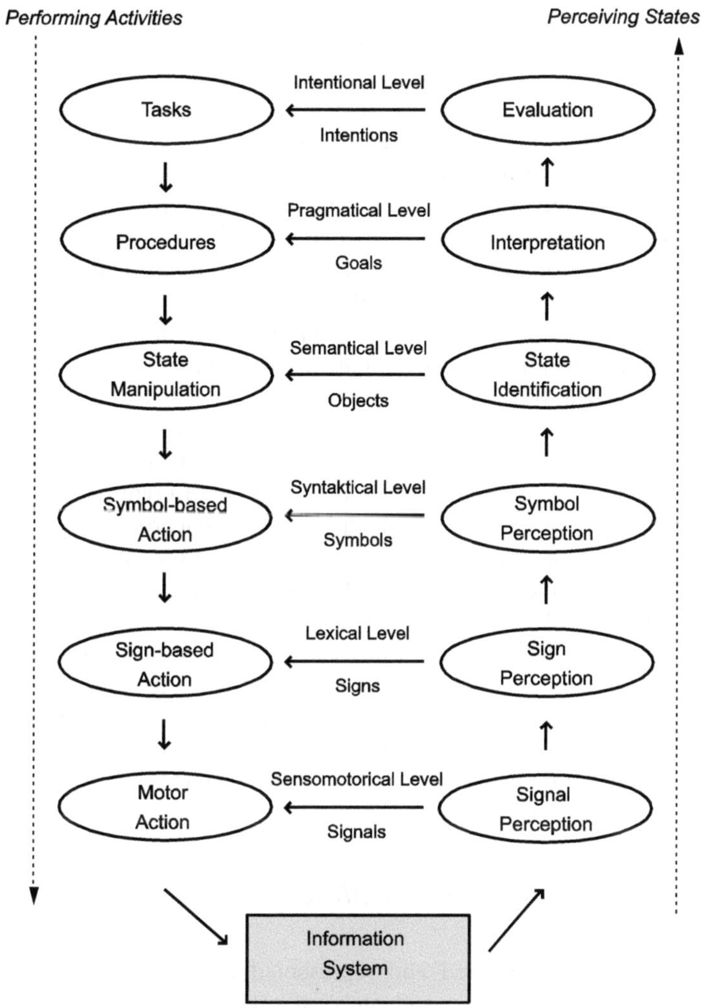

Fig. 6 Levels of interaction when performing a task (cf. Herczeg 1994, 2006a,c, 2009)

operator. The physical process itself might result in directly perceivable feedback in addition to the control system feedback. The operator will then perceive these feedbacks. Just as the activities had to be broken down from a task level down to the sensomotorical level, perception will take place vice versa (cf. Fig. 6 right side branch). This is quite similar to Rasmussen's decision ladder. One difference is a breakdown into the six levels as described for the breakdown of activities. Another difference is the observation that the operators set up expectations for system reactions of a certain kind on each level (anticipation). During the breakdown of activities they develop expectations of how the system should react according to their activities. These expectations will be used in guiding the perception process to check whether the activities have been successful. Any derivation from the expected outcomes might be reflected critically and can be a trigger for activity regulations on the level where the differences have been detected. As a result, errors may also occur because the operators are biased and focused on their expectations (Herczeg 2004). Quite similar to the decision ladder, the quality of the information presented to the operators will be vital in order for the regulations to be adequate and successful. An interesting part of this regulation process are the memory structures, the representations of knowledge and the information cues provided by the process control system to support these complex and sometimes fast regulations (real-time systems) on the different levels.

This model of stepwise refinement of high level tasks down to sensomotorical activities and the perception of system and environment feedback up the hierarchy can be transformed into Rasmussen's decision ladder when we start with system reactions (events), go up the perception hierarchy and back down the ladder of stepwise refinement of activities to cope with the situation. Therefore, both models seen from a psychological point of view are just two different perspectives of the same process. Tasks and events are just opposite sides of a process control situation. Tasks are refined into activities resulting in system reactions and events are system reactions to be perceived and translated into tasks. The information structures within these two models are basically the same. As a result, there is a good chance of designing one information system that can be used for task-oriented as well as event-oriented operations, which does not mean that both operational modes are the same. There are special operator needs to understand, plan and execute tasks as well as needs to recognize, interpret, evaluate and handle events. A common model will be part of the holistic model discussed later in this chapter.

2.6 Human Memory and Its Utilization

Information needs to be stored and represented in the human memory to be available for further processing and problem solving. Since the first models of human information processing there have been models describing the human memory and its characteristics.

An early groundwork of the understanding of the human memory was provided by George A. Miller's influential paper "*The Magical Number Seven, Plus or Minus Two: Some Limits on our Capacity for Processing Information*" (Miller 1956). His empirical work refers to a capacity limit of 7 ± 2 memory chunks of the human memory during perception of information and problem solving. This was the discovery of an "*immediate memory*" with certain access and coding properties. It was later called and conventionalized as the human *short term memory*. Additionally the abstract notion of a *chunk* as a memory unit leads to the remaining open question about the coding principles of human memory. However, in his 1956 paper Miller notes: "*[...] we are not very definite about what constitutes a chunk of information.*"

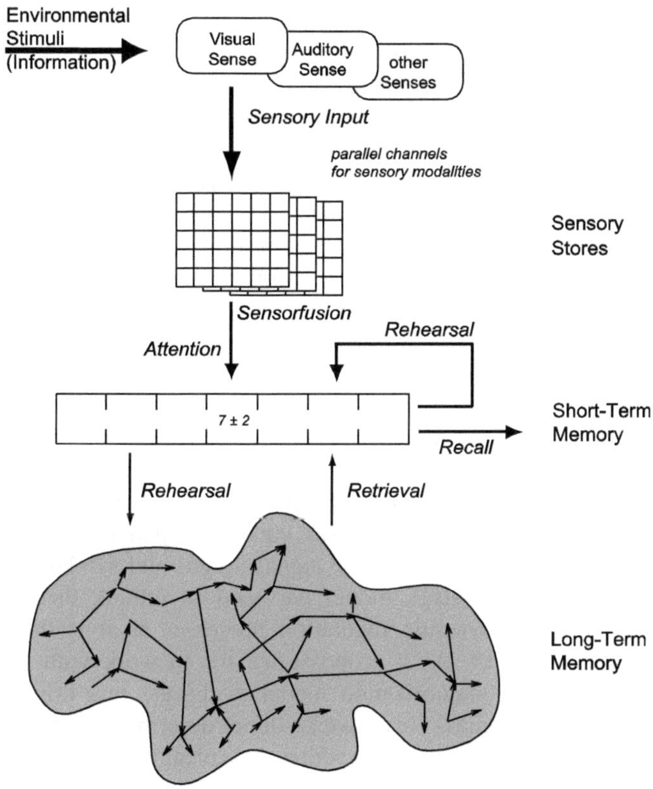

Fig. 7 Stages of human memory (cf. Herczeg 2009)

Miller's work and the later search for the characteristics or even properties of human information perception and information processing lead to models of *multi-staged memory structures*. Many researchers, such as Atkinson and Shiffrin (1968, 1971) as well as Card et al. (1983) developed similar multi-staged memory-models describing different types of memory usually starting with a sensory memory, feeding a short-term memory (STM) often called working memory, and collecting and structuring the memory chunks inside a so-called long-term memory (LTM). The memory stages are connected by memory mechanisms referred to as *attention*, *retrieval*, *rehearsal* and *recall* (cf. Fig. 7). Every memory stage or type has its own properties for access time, capacity, persistence or coding.

Generally, the value of this and similar models has been the insight that the human memory is not just an array of memory cells, but a highly differentiated active and dynamic structure of different memory types with specific performance characteristics. The differences lead to explanations why human memory succeeds or fails with respect to selectiveness, capacity, speed, accuracy or persistence. When it comes to the question of the memorization process in a scenario from information displayed in a cockpit to the recognition and storage of information, the limitations of the operator's memory become apparent.

By now, current neurobiological research indicates that there are many more memory types for different types of information (visual, auditory, tactile, spatial, temporal, etc.) and that there are quite complex memorization and learning processes within and between them (cf. Ware 2000). As an important step in understanding supervision and control activities of human operators we also need to assume a kind of transition memory for medium-term processes between short-term memory (working memory) and long-term memory (persistent knowledge storage). Some of these open questions about memory structures correspond to the difficulties describing simultaneous regulation processes in the activity hierarchy. The timing in this context ranges from a few milliseconds and seconds to hours and days of working on tasks and events.

A more refined memory model might result in more differentiation of the hierarchy of signals, signs and symbols. This refinement takes into account that there are obvious differences in coding of memory chunks as well as in the rehearsal methods as part of the regulation processes, e.g. visuo-spatial vs. phonological (cf. Baddeley 1999; Anderson 2000). While signals, signs and symbols fit quite well into Rasmussen's initial 3-stage model we need to rethink these models and implications for highly dynamic regulation processes as it is the case in the 6-stage model (cf. Fig. 6). As a proposal we try to work with the following 6-stage information and memorization model to discuss task-driven as well as event-driven control activities. This model will not describe all aspects of memorization or information processing, but it addresses typical information stages needed to explain or support task-based and event-based supervisory control activities (Sheridan 1987, 1988):

1. *Intentional Level*: goals and intentions
2. *Pragmatical Level*: symbolic phrases (semantic procedures based on logic)
3. *Semantical Level*: symbols (references and characterizations of meaningful objects)
4. *Syntactical Level*: lexical phrases (syntactic procedures based on grammars)
5. *Lexical Level*: signs and supersigns (basic coding of chunks)
6. *Sensomotorical Level*: signals and noise

In the field of general semiotics (Nöth 2000) and computational semiotics (Andersen 1997; Nake 2001) we can find descriptions and explanations for these types of activity-based memorization. With respect to information ergonomics we have to think about appropriate externalizations (e.g. presentations on displays) for these activity- and regulation-oriented forms of information. Ware (2000, p. 13) discusses PET- and MRI-based findings about the processing chain for visual symbols from perception to memory. Besides the multi-step neural processing he proposes a more abstract 2-stage model (Ware 2000, p. 26) with parallel processing to extract and match low-level properties of a visual scene as the first stage and sequential goal-directed processing as the second stage.

2.7 Search Strategies

One of the most frequent subtasks of operators is searching for information to accomplish a task or to cope with an event. The search strategies applied by the human operator will depend on:

- The structure of the information space;
- The size of the information space;
- The available keys (cues, input) for the search;
- The time available for the search;
- The quality requirements for the outcome of the search;
- The experience of the operator;
- The emotional state of the operator.

Searching for information in large information structures can be performed by applying different strategies (Herczeg 1994, 2006c). Basic *search strategies* are:

1. *Browsing*: The operator randomly searches information structures. In meshed information structures links between information nodes can be followed; in linear information structures a partially sequential search will be performed with more or less random jumps to other sequences of information (Fig. 8).
2. *Exploration*: The operator explores the information space more or less systematically. As a side effect of the search an overview (mental map) of the search space can be established. The map is a kind of meta-information (Fig. 9).
3. *Navigation*: In structured information spaces an operator may use a mental or physical map (meta-information) to guide the search and find the information needed generally faster than searching without the map (Fig. 10).
4. *Index-Based Search*: For a faster search an information space may be indexed, i.e. direct links leading from a list of keywords or a table of content to the information are available. The operator first looks up the information needed in the index and follows the link to the information itself from there (Fig. 11).
5. *Pattern-Based Search*: If the information needed cannot be indexed by precise keywords or a table of content, it might be looked up via diffuse search patterns (regular expressions) that are similar or somehow related semantically to the information. More or less intelligent search engines resolve the search pattern into references to information available in the information space and somehow related to the search pattern (Fig. 12).
6. *Query-Based Search*: In well defined and structured information spaces (information models) like databases, information search can be conducted using structured logic and algorithmic search specifications. Query languages and query interpreters resolve the query into references to the selected information contained in the information space (Fig. 13).

Information systems in safety-critical domains must often cope with the tradeoff between time required for the search and quality of the search results. It is one of the major challenges of the design of such systems to take this into account, i.e. to

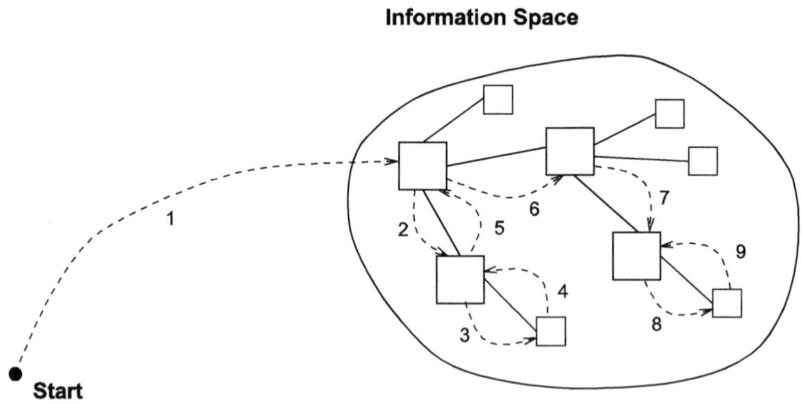

Fig. 8 Browsing

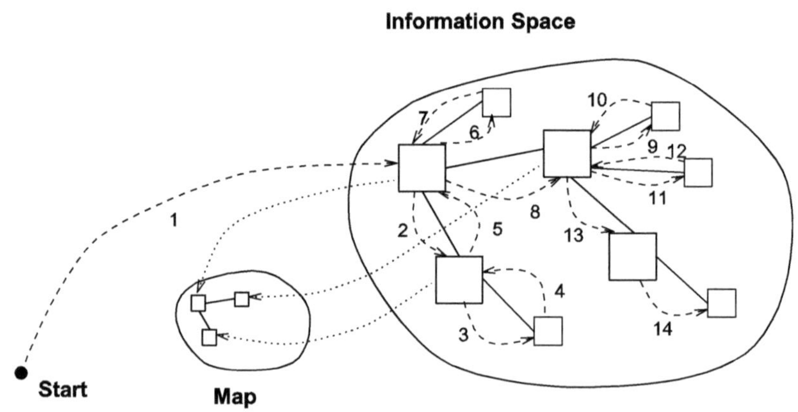

Fig. 9 Exploration

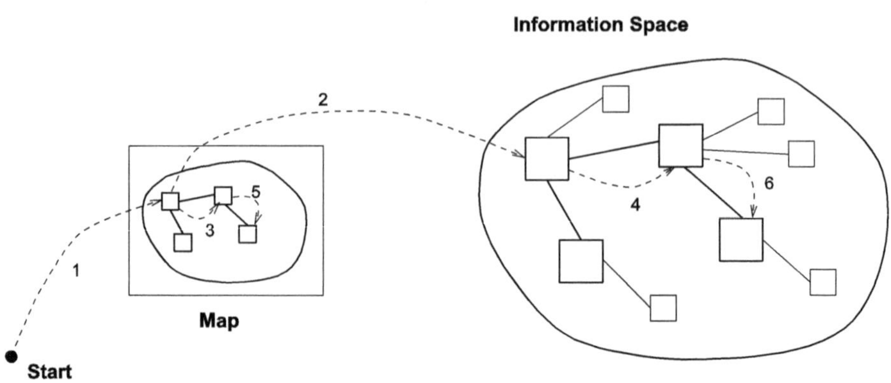

Fig. 10 Navigation

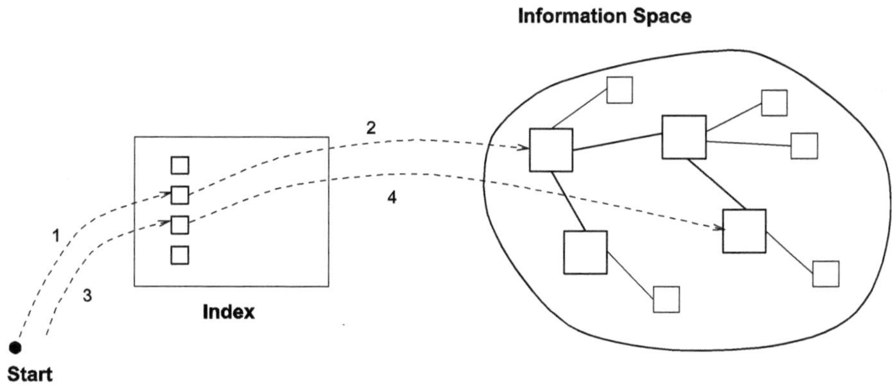

Fig. 11 Index-based search

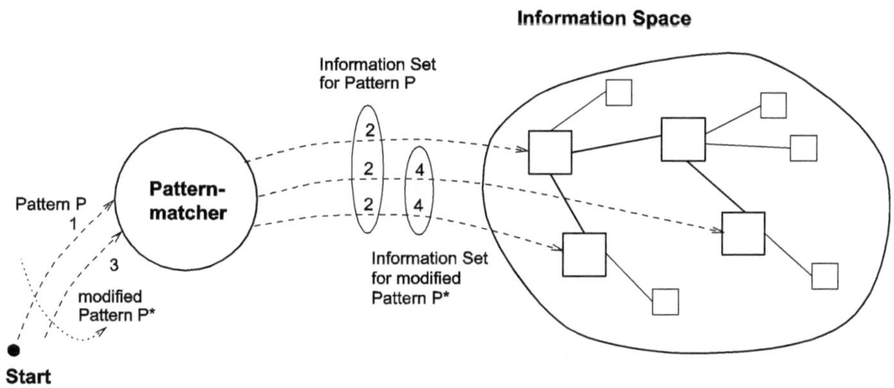

Fig. 12 Pattern-based search

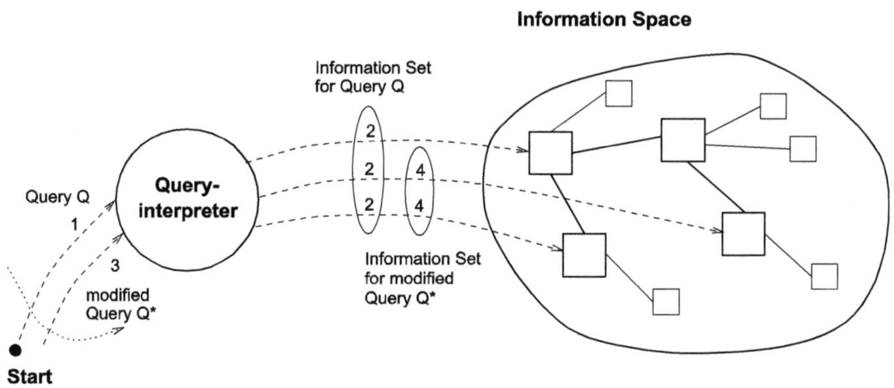

Fig. 13 Query-based search

provide search support which is time- and quality-sensitive with respect to the current operating context. As a pragmatic approach the provision of sufficiently appropriate results within the available time frame will be the rule. These have been called *satisficing solutions* (Simon 1981). Complete and best results will be the exception. For the operator it is important not only to be able to specify the search and obtain results, but to receive feedback about the progress of the search process and the quality of the results.

In many cases search will be related to the phenomenon of *serendipity*. Important information found is not always related to appropriate search strategies and search goals. In complex information systems many interesting and important facts will be found just by accident. Sometimes this disturbs the process of a systematic search and sometimes it results in satisficing and appropriate solutions.

Specifications of search, feedback for the ongoing search processes, and the presentation of the results will vary between the application domains, the experience of the operators and the contexts of operation. Much work has been done in the area of the visualization of results. Lesser work has been done in the area of progress and quality indications for search results and for other than visual information, like acoustic and haptic search feedback.

3 Limits of Current Cognitive Models

The cognitive models described can also be criticized with regard to their "*mechanistic design*" (cf. Dunckel 1986), even though they partly illustrate different aspects of action regulation. Crucial behaviour-determining constructs like emotion, motivation and processes of goal definition are not model-imminent or not differentiated sufficiently. In addition, there are no exact presentations and descriptions of operator's evaluation processes (preliminary evaluations) - becoming manifest in emotions – which in turn represent a driving force for the completion of actions and the initiation of subsequent actions (Stein 2008).

A further point of criticism is the memory of action-relevant knowledge, since Rasmussen (1983, 1986), for instance, did not reflect on a memory model or cognitive processing (cf. Dunckel 1986). It must be assumed, however, that certain action patterns are stored over a prolonged period of time to be retrieved when they are needed.

Furthermore, it should be noted that social processes and interactions or also organizational processes – in which action is usually integrated – are not detailed and referred to in the models described (cf. Cranach and Kalbermatten 1982).

Most of the cognitive activity models described in the previous sections are characterized by a clear separation and discrimination of the subprocesses. The problem with these kinds of models is that we can also observe certain effects that cannot be associated with these models, e.g.:

- Operators do not proceed systematically in the order of these levels; instead they jump between model levels, omitting certain steps (cf. the shortcuts of Rasmussen's decision ladder);
- Operators developmental models (e.g. Carroll and Olsen 1988) that enable them to mix model levels and show performances that are more holistic than substructured; for example the mixture of syntactic and semantic information processing enables human operators to disambiguate perceived signs to symbols (Herczeg 2006b, c);
- Cognitive and emotional models need to be combined to explain certain human behaviour in control tasks; historically cognitive and emotional models have been discussed and analyzed mainly separately;
- Human memory models have to be redefined through current findings about brain structures and neural processing; for example the idea of a staged memory (e.g. Miller 1956; Atkinson and Shiffrin 1968, 1971; Wickens and Hollands 2000) does not describe memorization or learning processes very well;
- Multiple and parallel activities can only be partially understood by applying the available activity and the resource models (e.g. Wickens and Hollands 2000); we need a more complex combination of these models which describe the multiprocessing abilities of human operators more appropriately.

Humans can perform several activities at the same time. This is sometimes called *multitasking*. Some of the activities take place consciously, but many of them are performed subconsciously, i.e. we are only partially aware of them. The discussion of activity models (cf. Fig. 6) revealed different levels of activities and regulation processes. Activities on a higher level are refined into more fine-grained activities on the next lower level. Low-level perceptions are analyzed and enriched to some fewer higher perceptions. If the perceived signals, signs or symbols do not correspond to those which have been anticipated, regulation may be applied. As a result of this activity refinement and perceptual aggregation we will have many parallel regulation processes proceeding on different levels, some of them consciously (mainly the upper levels) and some of them subconsciously (mainly the lower levels). Additionally, several dependent or independent tasks can be started and performed in parallel.

How and to what extent will an operator be able to follow and control tasks which are performed simultaneously? The models available for describing the resources and capacities to perform these processes (e.g. Wickens and Hollands 2000; Endsley 2000; Herczeg 2009) as well as the resulting load, strain and stress do not correspond in every respect to the complexity, the memory effects and time structure of such simultaneous activities. The current models can mainly be used to discuss coarse-grained effects of multitasking. This affects the design of information systems for real-time contexts like transport. How can we optimize information displays when we do not have a fine-grained understanding of multiple activities in real-time situations? The approach that has been used is to give the operators any information they might need, observing the overall performance and the standard errors occurring. This results for example in cockpits which are filled with controls and displays, many of them multi-functional, leaving the usage and adaptation to the operators and their capabilities. More context- and task-sensitive environments could provide better selective support of operations by presenting the main information needed, by checking time conditions to be met and by controlling the performance of the operators. The concept of *Intention-based Supervisory Control* (Herczeg 2002) proposes smarter process control environments, taking into account which activity threads are ongoing between a human operator and a semi-automatic control environment.

Models and automation concepts for multiple parallel activities should be discussed and developed with respect to multiple operators and multi-layer automation (Billings 1997) to optimize the dynamic real-time allocation of tasks between human and machine operation.

The following shall outline an extended model for describing action regulation during access to an information system, observing and acknowledging the described strengths and weaknesses of the already presented models. The design of this model is based on the presented models, however, in particular motivational and emotional processes, the formation of expectation as well as expectation-value-comparisons during access to an information system are modelled. To this end, all relevant constructs, e.g. motivation and emotion, cognitive resources and workload, effort regulation, fatigue and stress are initially described and subsequently put into an overall context for the field of information systems in transportation.

3.1 Motivation and Emotion

The intention of motivation psychology is to "*(...) explain the direction, persistence, and intensity of goal-directed behaviour. (...) the two main, universal characteristics of motivated behaviour are control striving and the organization of action*" (Heckhausen and Heckhausen 2008, p. 3; Schneider and Schmalt 2000). In this context Heckhausen and Heckhausen (2008) differentiate between goal engagement and disengagement. The former involves focusing (attention, perception) on triggering stimuli, the fading out of goal- and action-irrelevant stimuli and the "provision" of partial actions. Therefore the expectation of the effectiveness of the planned behaviour is optimistic. In contrast, goal disengagement can result when the expected effectiveness exceeds the costs of behaviour. This includes a devaluation of the endeavored goal and the enhancement of the value of alternative goals. The processes of goal engagement and disengagement play a major role in the use of information systems. This is due to the fact that information access happens in a dynamic environment and thus priorities must be assigned to several competing information goals. Furthermore, depending on changing environmental, process and system states a decision must be made during the search process on whether to continue the information search or discontinue the search in favour of other information goals or also other activities. In order to be able to model the motivational aspects of the access to an information system a general motivation

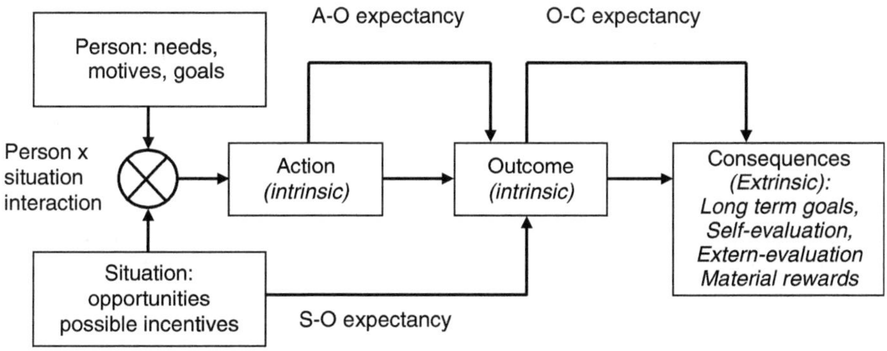

Fig. 14 A general model to describe the determinants of motivated action (Heckhausen and Heckhausen 2008)

model by Heckhausen and Heckhausen (2008) can be used as a basis to describe the determinants and the process of goal-directed behaviour.

As Fig. 14 shows, the motivation of an individual is determined by the interaction between personal and situational factors as well as anticipated outcomes and longer term consequences. With regard to personal factor described in the model, Heckhausen and Heckhausen (2008, p. 3) differentiate between:

- Universal behavioural tendencies and needs (e.g. hunger, thirst),
- motive dispositions (implicit motives) that distinguish between individuals, and
- the goals (explicit motives) that individuals adopt and pursue.

For the use of information systems explicit motives (self image, values, and goals) are relevant because they can be described as the superior layer for achieving information goals. However, motives like curiosity (particularly with regard to complex systems which the operator cannot cover by cognitive means), achievement motivation (implicit motives), and safety are also of great importance. In flight simulators, for instance, we can observe pilots exploratively investigating information systems. Since this is also the case for system-experienced pilots we can draw the conclusion that the complexity of information systems cannot entirely be covered cognitively, leading to a reduction of insecurity by means of explorative investigation (curiosity).

Apart from the personal factors already described which contribute to motivational tendency certain situations implicate cues perceived by individuals as possible incentives or opportunities. These stimuli which may have positive or negative connotations are associated with action, the outcome of action and also the consequences of action. With regard to situations Heckhausen and Heckhausen (2008) differentiate between situation-outcome-expectancy, action-outcome-expectancy and outcome-consequences-expectancy. The first-mentioned expectancy means that results will come to pass even without performing own actions – i.e. situation-dependently – and thus initiation and performance of action is improbable. In contrast, goal-oriented action is performed in case situation-outcome-expectations

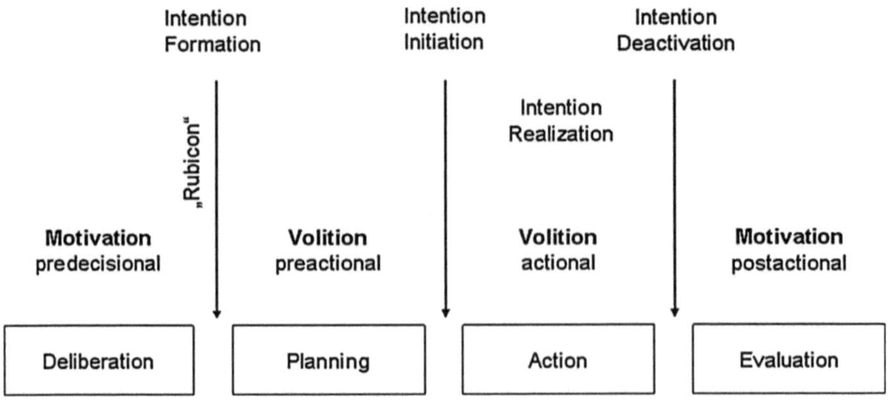

Fig. 15 The Rubicon model of action phases by Heckhausen and Gollwitzer (1987)

are low, while at the same time action-outcome-expectations and outcome-consequences-expectations are high.

If congruency is obtained between the characteristics of a motive and the possible incentives a motivational tendency will result. This tendency, however, is not sufficient for an individual to "feel" closely connected to action goals or for goal-oriented action. Heckhausen and Heckhausen (2008) explain that intention should emerge from a motivational tendency. They illustrate the individual phases from the formation of intention to the action and the evaluation of the action in their Rubicon model. Figure 15 shows that they differentiate between motivation (predecisional and postactional) and volition (preactional and actional).

The deliberation phase (predecisional) is characterized by wishes and contemplation of pros and cons concerning potential goals (Heckhausen 1989). In this context, especially the assessment of the feasibility of one's goals in terms of practicability with regard to the context or environment of action and one's own capabilities play a decisive role. The anticipated implementation of the goals is also evaluated. "*The expected value depends on how disagreeable or agreeable the potential short- or long-term consequences and their probability of occurrence are estimated to be*" (Gollwitzer 1996, p. 534). Since the process of deliberation is not continued indefinitely Heckhausen (1989) assumes a "*facit tendency (concluding)*". This tendency implies that the more comprehensively a person has deliberated the information available for a decision and the longer the process of deliberation takes the closer he or she perceives the act of making a decision. When a decision about an action alternative has been made, leading to an intention, the person feels committed to achieving the goal (Gollwitzer 1996; Heckhausen 1989). Heckhausen and Gollwitzer (1987) call this process "*crossing the Rubicon*".

In the planning phase (preactional, volition) the focus is on the schedule of actions to achieve one's goals. According to Gollwitzer (1996, p. 536) "*primarily the 'when' and 'where' of the initiation of action as well as the 'how' and 'how long' of the progress of action is dealt with in this phase with the objective of committing oneself to one of the many possibilities, initiating, performing and completing goal-implementing action*". Whether an intention is implemented into action and realized depends on the volition strength of the intention in relation to the volition strength of other, competing intentions and on the evaluation of the action context (favorable, unfavorable) with regard to initiation. Heckhausen (1989) points out that usually many intentions which have not been implemented are on hold during the preactional volition phase in a sort of waiting loop, "*competing with each other for access to the action phase*". In this phase a certain action can, for instance, be postponed due to an unfavorable context and more favorable action alternatives may be performed for the time being.

According to Gollwitzer (1996, p. 537) action initiation is the "*transition to the actional phase. Goal-directed action is characteristic of this phase*". Action control is implemented by means of mental models and goal presentations (Heckhausen 1989; cf. Miller et al. 1960; Volpert 1974; Hacker 1978, 1986). Goal achievement is evaluated in the postactional motivation phase. One of the questions arising in this phase is "*whether the actual value of the achieved goal corresponds to the expected value*" (Gollwitzer 1996, p. 538). These evaluations generally result in a more adequate assessment of the value of a goal intention in future decision-making processes in the predecisional phase.

When observing the process of accessing an information system as well as the reception and cognitive processing of information, i.e. the implementation of information into a process control task, the following motivational phases illustrated in Fig. 16 can be differentiated:

- Motivation for information system access,
- Maintaining motivation within the information system,
- Appraisal of informational content (with regard to the process control task),
- Motivation for implementation,
- Outcomes of implementation, and
- Appraisal of the total "information" process (Stein 2008; Stein et al. 2002).

With reference to the general motivation model by Heckhausen and Heckhausen (2008) described above, personal and situation-related factors play a behaviour-initiating and determining role with regard to accessing information systems (motivation for information system access). This is the case, for example in aviation, when a pilot accesses an information system (e.g. electronic flight bag). In this context the determinants for a motivational tendency are goals (e.g. safe control of an aircraft, accuracy) and goal prioritization (e.g. safety is more important than personal needs) as well as the expectations of the operator (e.g. consulting an information system might be helpful in bad weather conditions) on the one hand and situational factors, e.g. a certain system state (e.g. engine failure) which is indicated on a display on the other. After a motivational tendency has become a goal-directed intention and goal-directed action (information system access), motivation must be maintained in the "information system space" and during the search for information.

After the information has been found and cognitively processed it is integrated into process control after its relevance has been considered. Subsequently, the total

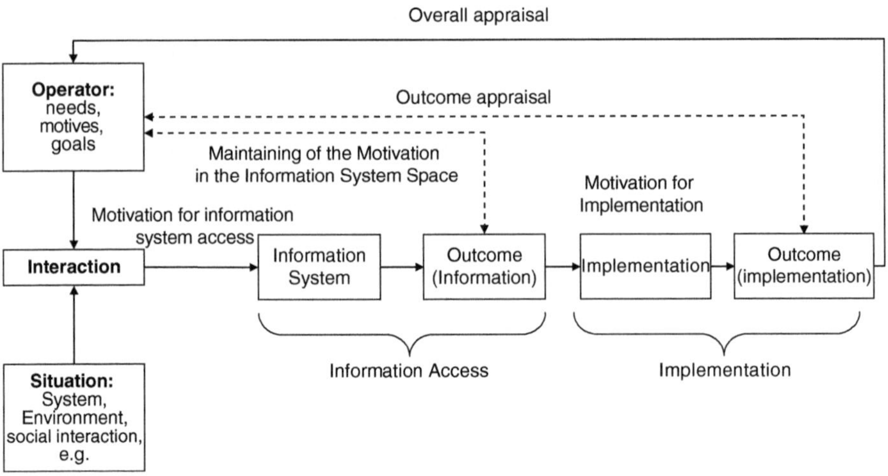

Fig. 16 Motivational phases when accessing an information system and implementing information (Stein 2002, 2008; Stein and Müller 2005)

process starting with information access up to the implementation is evaluated by the operator. The built-up expectations (anticipated information) are compared to the result and evaluated.

Expectations with regard to scope, quality of information as well as handling of the information system play a major role for the emotional processes which represent the evaluation results of the use of the information system as well as the received information (cf. Stein 2002; cf. Sanders 1983). In this case expectations are built up from the information stored in long-term memory which are compared to the "monitor outputs" (information system) in a continuous process, leading to new evaluations (emotions). In turn, these are stored in long-term memory, affecting perception as well as cognition and action selection/implementation.

Emotion is a construct which is closely linked to motivation. In this respect Scheffer and Heckhausen (2008, p. 58–59) state: "*emotions can be described as a rudimentary motive system that serves the internal and external communication of motivational sequences. (…) Emotions are thus prerational forms of values and expectancies that influence the motivational process.*"

Immelmann et al. (1988) define emotion as "*physical, intellectual and emotional reactions of an individual to environmental occurrences which are significant to his own needs. This emotional evaluation of a situation often leads to a clear sense of pleasure or aversion, physiological arousal changes and specific action tendencies*" (Immelmann et al. 1988, p. 878; cf. Schneider and Schmalt 2000).

Schneider and Dittrich (1990, p. 41) specify three characteristics of emotion:

- "*Specific ways of experience*" (only perceived by the individual),
- "*Motoric behaviour*" (e.g. expressional behaviour), and
- "*Peripheral physical changes*" (e.g. change in blood pressure, transpiration, etc.).

According to Grandjean et al. (2008) three general approaches to the modeling of emotional mechanisms can be distinguished in affective science:

- Basic emotion models (basic emotions are elicited by certain stimuli, resulting in prototype facial expression, physiological changes as well as certain action tendencies; the various scientific approaches differ in the quantity of assumed basic emotions (e.g. fear, joy, sadness)),
- Dimensional feeling models (feelings are described in the dimensions of arousal, valence and potency, etc.), and
- Componential appraisal models "*emotions as a dynamic episode in the life of an organism that involves a process of continuous change in all of its subsystems [e.g., cognition, motivation, physiological reaction, motor expression, and feeling – the components of emotion] to adapt flexibly to events of high relevance and potentially important consequences*" (Grandjean et al. 2008, p. 485).
- Neurobiological models (emotions are a product of stimuli categorization either via limbic system ("quick and dirty") or via reappraisal in the prefrontal cortex, see Pinel, 1993).

Scherer (1990, p. 3) postulates that "*emotions represent processes, each of which involve various reaction components or modalities*". In his Component Process Model Scherer (1984) establishes four phases of sequential assessment:

- *Relevance* (information is assessed for novelty and entropy),
- *Implications* (consequences and the impact of the information concerning long-term goals and well-being),
- *Coping* (strategies to cope with the consequences of the information),
- *Normative significance* (impact on self-concept and social norms).

Within the scope of information system access and implementation those emotions in particular are significant which the recipient feels and perceives during:

- Information access (e.g. curiosity or the hope of finding the information required quickly and within a reasonable period of time),
- Information search (e.g. annoyance if the path leading to the information is impaired by non-adequate navigational aids or joy after finding the required information),
- Cognitive information processing (e.g. frustration if information is not understood or if high cognitive effort is required to understand the information), and
- at the end of the information retrieval (in terms of an evaluation of the sub- or total process on the basis of which future decisions on information access will be made).

Emotions play a major role for information access, particularly with regard to the operation of complex systems (aircraft, ship, etc.). Thus, certain emotions, for example fear in combat missions with fighter aircraft, can adversely affect cognitive abilities. Consider a fearful pilot who handles a checklist superficially, thus makes a wrong aeronautical decision and misses an enemy on his radar. By this means he might lose a combat mission.

3.2 Cognitive Resources

Common single and multiple resource theories (Wickens 1980, 1984; Norman and Bobrow 1975; Kahneman 1973) are based on a metaphor describing "*pools of energy*", or resources provided and exhausted during task performance (Szalma and Hancock 2002, p. 1). With regard to the analogy used, referring to the field of economics or thermodynamics/hydraulics, Szalma and Hancock (2002) state that it is not unproblematic to choose a nonbiologic metaphor for a biological system since it can generally illustrate the complexity of living, complex and dynamic systems only to a limited extent.

Early, so-called single resource theories (Kahneman 1973; Treisman 1969 etc.) postulate a singular pool with limited resources/capacities by means of which a multitude of different tasks can be performed. The quality of performance depends on the task demands and the degree of availability of the resources. However, Kahneman (1973) already assumes that cognitive resources can be characterized as dynamic and they are thus subject to certain variability. In this context Gaillard (1993) adds that besides interindividual differences regarding processing capability between operators there are also intraindividual differences depending, for instance, on sleep loss, anxiety or also psychosomatic complaints (cf. Fairclough et al. 2006).

Norman and Bobrow (1975) broadened the single resource theory by Kahneman (1973) to include aspects of resource-limited and data-limited processes. With regard to resource-limited processes a higher degree of effort (also greater exhaustion of resources) leads to improved performance, while in case of data-limited processes performance is dependent on the quality of data (information) with respect to data-limited processes and a higher degree of effort does not contribute to improved performance. Irrespective of the cited differentiation by Kahneman (1973) the most important disadvantage of the single resource theories is that these cannot explain how highly trained operators can manage time sharing of multiple different tasks.

This is where multiple resource theories apply (Navon and Gopher 1979) with humans being considered with multiple information processors. Each channel has its own capacities and resources which can be distributed to different tasks and do not interfere with each other. The Multiple Resource Model by Wickens (1980, 1984) follows this tradition and also represents an explanatory approach regarding the possibility of parallel processing of tasks. Wickens differentiates between four dimensions of independent resources:

- Stages of processing (perceptual, central and response),
- Codes of processing (verbal, spatial), and
- Modalities of the input (visual and auditory) and output (speech and manual).
- Responses (manual, spatial, vocal, verbal)

According to this Multiple Resource Model the cognitive workload increases when the same resources need to be utilized for two tasks to be performed in parallel (primary and secondary task). Task demand, resource overlap and allocation policy are workload determinants within the scope of the Multiple Resource Model (Wickens and Holland 2000).

3.3 Workload

With regard to the correlation between the workload concept and the Multiple Resource Model, Wickens (2008) explains that both constructs overlap in terms of their contents, although they are separate concepts. With respect to differentiation he further explains that the workload concept is closest related to task demand. However, it is less linked to resource overlap and allocation policy.

This can also be illustrated by a workload definition by Hart and Staveland (1988) in which the authors focus on the correlation between task demands and the required or available cognitive resources the relation of which then determines the extent of workload. "*The perceived relationship between the amount of mental processing capability or resources and the amount required by the task*".

Hart and Wickens (1990, p. 258) define workload as a "general term used to describe the cost of accomplishing task requirements for the human elements of man–machine-interface". In this regard the costs can be discomfort, fatigue, monotony and also other physiological reactions.

With regard to task demand Wickens and Holland (2000) differentiated between "*two regions of the task*". In the first region task demand is lower than the resources of the operator. Thus, the operator has residual capacity even in critical or unexpected situations to maintain action regulation. In the second region the demands exceed resources (high workload), resulting in a performance breakdown (Wickens and Hollands 2000).

The significance of the workload concept in the context of user interface design and information systems can be illustrated by the following items by Gaillard (1993):

- When the operator works at the limit of his cognitive resources the occurrence of errors will be more likely;
- Underload can also make the occurrence of errors more likely;
- When the operator works at the limit of his cognitive resources for a longer period of time, stress reactions may result.

There are subjective (workload rating scales) and objective methods (secondary task method, physiological measurement techniques (cf. NATO Guidelines on Human Engineering Testing and Evaluation 2001) which can be conducted to measure workload. When the reasons for high workload have been identified in a workload assessment, the interface design or the task can be modified.

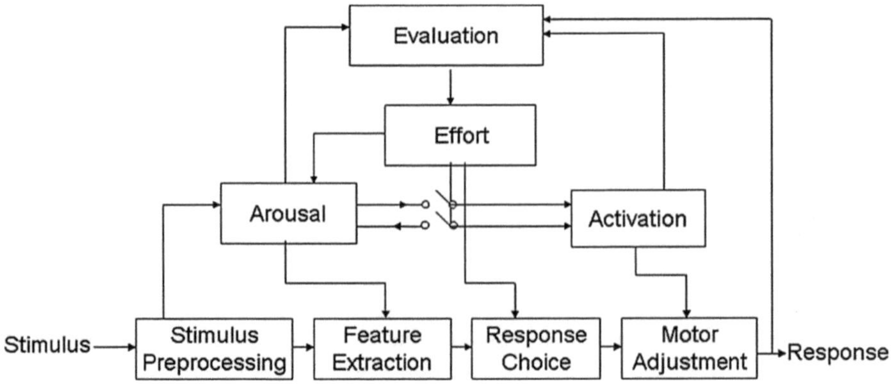

Fig. 17 Model of energetic mechanism (Sanders 1983)

3.4 Effort Regulation

In some situations greater effort is required for achieving goals. It is thus possible when using information systems that an information goal cannot be achieved with the effort anticipated by the operator. In this case the operator must decide, depending on the current status of the transportation system and also further environmental factors, whether the effort should be enhanced, the goal should be adapted lower or the information task should be discontinued. Such a decision is generally justified by multiple causes. However, as already mentioned above, an increase of effort can only contribute to an improvement of achievement in resource-limited processes.

Sanders (1983) describes in his model of energetic mechanism the relations between human information processing stages and an energetical system (Johnson and Proctor 2003). In his model he suggests that there are three energetic mechanisms (arousal, effort and activation) allocated to feature extraction, response choice and motor adjustment (cf. Fig. 17). In this context arousal triggers feature extraction, effort triggers response choice and activation triggers motor adjustment. If there are differences between the intended and the obtained response, an evaluation mechanism stimulates the effort level which in turn increases arousal and activation, thus intensifying information processing. If the intended and the obtained result correspond in the evaluation there will be no increase of effort.

When using information systems goals can be achieved also, for instance in case of usage problems, if the operator does not find certain information directly, by means of an increase of effort. If the user is, for example, overwhelmed by the diversity of information or also navigation options (cognitive overload), an increase of effort generally does not result in the desired effects.

3.5 Stress and Fatigue

Stress and fatigue are two constructs which are rarely integral parts of the common action regulation theory. At the same time, the influence of these constructs on the quality of performance and safety, particularly in critical or unpredictable situations, is very significant (cf. Hancock and Szalma 2008; Stein 2008). Particularly in aviation information systems are accessed in critical situations, stress and fatigue having an adverse impact on attention and also on cognitive information processing.

Gaillard (1993, p. 991) defines stress "*(...) as a state in which the equilibrium between cognitive and energetical processes is disturbed by ineffective energy mobilization and negative emotions. Stress typically is characterized by ineffective behaviour, overreactivity and the incapacity to recover from work*".

Hancock and Szalma (2008) further explain that all stress theories are based on the assumption of an evaluation mechanism which is used to evaluate events with regard to their psychological and physiological relevance (well-being) for the organism. In addition, it is assumed that individuals control internal energetic processes to compensate equilibrium imbalances, e.g. caused by task demands.

The best-known and most elaborated stress theory is published by Lazarus (1974, cf. Lazarus and Folkman 1984). He assumes that there are three levels of event evaluation:

1. Primary appraisal (evaluation of whether an event has positive, non-relevant or jeopardizing effects on the organism. The evaluation of "*jeopardizing*" is differentiated still further as "*challenge*", "*threat*", "*harm*"/"*loss*").
2. Secondary appraisal (check if there are sufficient internal resources to fulfill requirements. If internal resources are not sufficient, stress reactions will occur. In this context coping strategies will be developed to be able to handle the situation).
3. Reappraisal (the total event is re-evaluated due to the previous appraisal steps).

With regard to the use of information systems stress occurs if information required for main controls in an emergency situation, for instance, is not available or not available as rapidly as required or if the information does not offer adequate support to solve the problem or an adequate solution approach. Generally, stress is associated with negative emotions (e.g. anxiety, anger, irritation).

Furthermore, stress reactions are accompanied by the following states:

- The operator has problems concentrating on the task,
- More energy is mobilized than is required for the task,
- The energetic state is not adequate with regard to goal achievement, and
- The activation level of the operator remains high, even after the task has been accomplished; the operator has problems returning back to a normal activation level (Gaillard 1993).

Fatigue occurs in particular in vigilance situations in which the operator for instance fulfills monitoring tasks and in which "signals", which must be observed, are rare events (e.g. during certain flight phases during which the pilot must fulfill only monitoring tasks). In general, the workload during such situations is very low (c.f. Hockey 1986) and is accompanied by a feeling of monotony and fatigue. If information is then urgently required, for instance from the electronic flight bag, the activation level and also its resources are low, so that the organism is not able to react quickly to the requirements.

With regard to the construct of fatigue, Boff and Lincoln (1988) declare that according to the definitions fatigue is frequently associated with a tired feeling, however, an operational definition is difficult. Fatigue is frequently defined as the number of hours of work or hours on duty, although it depends on numerous internal and external factors of each person when fatigue sets in and when it does not. Apart from vigilance situations fatigue can also result from a prolonged high extent of information overload (Hockey 1986).

4 Regulation of Activities in an Information Task Environment

Prior to the access to an information system the operator has to identify the requirement and necessity of accessing additional information from an information system based on current system states (e.g. aircraft), environmental or also person internal cues. If, for instance, a flight has to be redirected at the destination airport on short notice due to adverse weather conditions (environmental cues) the pilot will initially recognize the need to find information on the alternative airfield using the electronic flight bag, if he is not familiar with this particular airport. The same is true for the failure of a subsystem, e.g. the hydraulic pressure system (system cues); also, if one of the pilots falls ill or experiences fatigue during a flight and is thus no longer able to fulfil his task fully or not at all the procedure for such a case which is stored in the electronic flight bag will be accessed (person internal cues). The question of whether these cues are perceived and to what degree of accuracy they are perceived generally depends on the competence and experience and also on the self-perception and subjective values of the operators with regard to process control and the information system. These perception processes, i.e. perceiving relevant cues and recognizing and deriving an information task on that basis, constitute the first level of situational awareness. In terms of psychology of motivation these cues serve as opportunities and possible incentives, resulting in motivated behaviour (access to an information system) in the interaction with needs, motives and goals, depending on how pronounced these are; this is described in Chapter "Motivation and Emotion" (cf. Fig. 18).

According to Rasmussen (1983) information tasks without cognitive parts can be processed also on a skill-based level following a cue (skill-based level). In the

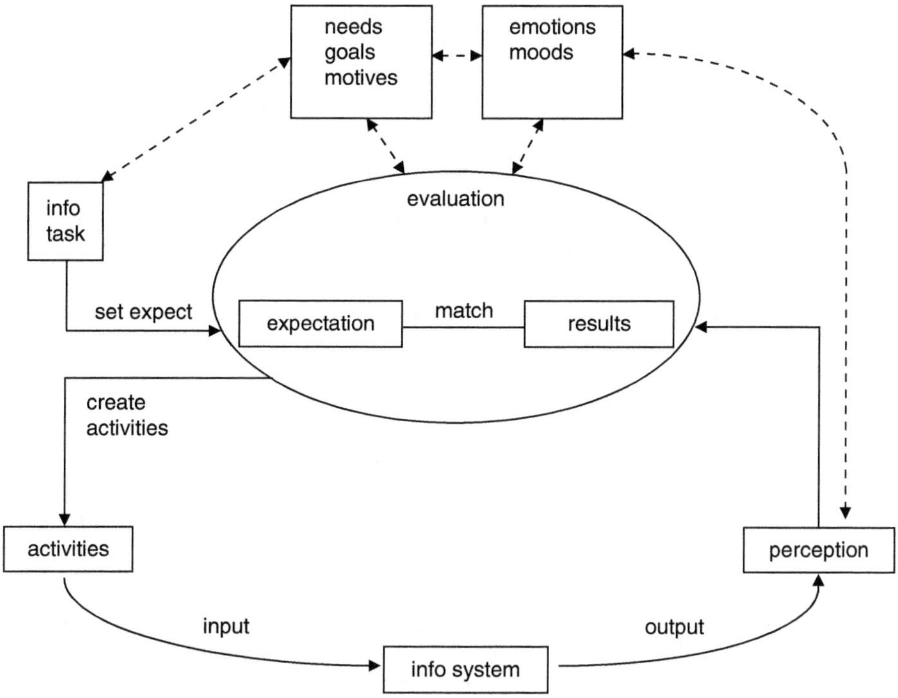

Fig. 18 Model of human–information-system–interaction

aviation sector these so-called habit patterns (stimulus x = reaction y) are trained specifically to enable the safe and constant execution of a certain behaviour also in high-stress situations. This also comprises the access to specific information following certain relevant cues.

However, at this point it is possible that the information task cannot be "accepted" and performed due to prioritization and also due to a reluctance to do so aversion as (emotion). In the aviation sector, for instance, a pilot may experience reluctance if he has already processed a checklist (in an electronic flight bag) repeatedly during several short-distance flights on the same day, thus deciding not to complete the checklist during the next short-distance flight. In addition, decision processes may occur in which the pilot decides not to access an information system since the required period of time for the information to become available is too long and the access thus not beneficial to process control. Furthermore, prioritization may alter during the access to an information system due to system or environmental changes, resulting in the termination of the information search.

The formation of expectations happens as part of the interaction between personal characteristics (abilities and skills, subjective formation of values, etc.), information goals (also with regard to accuracy and clarity of information goals), priorities within process control as well as the perceived characteristics of an information system (configuration, information organization, navigational aids, stringency, etc.). On the one hand, expectations are generated by practise, training and application and can be explicit or implicit. They are based on experiences with certain information systems (or information system categories):

- Similar information systems which are applied to the current application,
- The information paths offered by the information system and the operator's own search strategies (duration and also "ease" of the information search),
- Quality, media type (image, text, video, hypermedia etc.) and scope (degree of detail) of information, and
- Practical suitability and context sensitivity of information (with regard to information application).

On the other hand, expectations regarding the configuration and function of an information system may be based on hearsay (e.g. "heard from another driver how complex and difficult it is to use the driver information system of a certain brand") and more or less realistic mental models.

Regardless of whether expectations are based on a realistic model of an information system or information characteristics they establish the basis of an expectation-result appraisal which is made not only at the beginning and at the end of the information system access, but also in iterative steps during the search process.

In the transport sector expectations are broadly diversified on the realistic-vs.-unrealistic scale due to the differences regarding professionalism (e.g. average pilot vs. average driver). Pilots are trained to access an information system in the cockpit in type rating training and familiarization. This automatically results in the formation of realistic models and expectations which are generally confirmed automatically in an expectation-result comparison. In contrast, the automobile sector encompasses the total range of professional and unprofessional drivers (wide population distribution). In terms of information system design this results in special requirements for the adaptation of abilities and skills.

If the operator has decided to access an information system he will in a professional environment generally revert to strategies he has learned in order to achieve his intended information goals. These strategies may be well trained search algorithms (navigation, query-based search) or also trial-and-error strategies (browsing, exploration, pattern-based search) if the operator has little experience with the information system. The latter can frequently be observed also in modern flight decks with a high level of automation. This automation complexity (and also information presentation complexity) often exceeds the cognitive resources of the operator.

The achieved information results are initially processed cognitively. This means that they are integrated in existing knowledge, i.e. working memory (in case of high relevance and high probability of repeated use they are stored in the long term memory). Subsequently, the final expectation-result-comparison is initiated.

The required information may not be found promptly or only fragments of the anticipated information may be available or scope and degree of detail may not fulfil the expectations of the operator. Thus, cognitive processes subsequently take place during which the operator calculates the usefulness of information for process control vs. costs or time and then decides whether the search for information is continued or discontinued. Furthermore, if information goals are not achieved regulation processes take place during which either information task requirements (requirement level) are reduced or mental effort is increased (mental effort, cf. Sanders 1983; see also action regulation, cf. Herczeg 2009). Greater mental effort on the human energetic level – as described by Sanders (1983) – affects arousal (effectiveness of recognition of stimuli and dealing with stimuli) on the one hand and activation (improvement of motor output) on the other.

It should also be noted that in addition to the goal expectations of the operator environmental or system factors which are partly the basis of "formulated" goal expectations may also change in the course of a search process. This dynamic means that there may be a certain information need at a certain point of time X during the use of an information system; this need, however, is postponed prior to the completion of the information search and cognitive processing due to a new (more important) information requirement. For this reason, especially aspects of prioritization of information and also goal expectations play a major role for the user. Moreover, it can be assumed that the extent of the scope and quality or the degree of detail of the goal – i.e. the information required – which is known to the user of an information system varies and that this degree of detail can also change with the movement within the information space.

After the conclusion of the comparison described between the expectations formed and the results (or perceived results) achieved these are weighted in terms of emotion. This means that the entire informational process is associated with a feeling of frustration or even anger if for instance the information goal was not achieved. This emotion will determine the probability of access with regard to the next intended access, controlling the motivational processes. If a certain behaviour (information access) is successful in a specific situation this behaviour is repeated as required and also generalized for other situations. If a certain behaviour is not successful and associated with negative emotions the probability of repeated execution is low.

5 Summary and Conclusions

Common action regulation models were used initially to develop a model describing Human-Information-System-Interaction. The analysis of the models indicated that the focus is almost exclusively on perceptive and cognitive processes and that other relevant aspects, e.g. emotion and motivation, also workload and effort regulation are only partly or implicitly integrated in these models. To set the theoretical stage for an extended model constructs like situational awareness, workload, fatigue and decision making were presented in context and subsequently integrated in the Human-Information-System-Interaction model. The approach consists of cognitive, motivational as well as emotional processes during access to an information system, information search, information access, the decision making process as well as the resulting action (cf. Stein 2008). The background can serve as a basis for the design and configuration of information systems in various areas, in particular to develop a comprehensive focus on the operator side. Furthermore, the model can provide a theoretical framework for ergonomic analyses of information systems and displays. This is of particular significance in all transportation sectors since the development of electronic flight bags or driver information systems, for instance, has been based exclusively on experience and hardly on theoretical principles. However, theoretical principles form the basis of user-centered development and the systematic advancement of these methods. Finally, it should be mentioned that at present, it is difficult to combine and synthesize the theoretical constructs to one superior and empirically validated model (see also Tsang & Vidulich, 2003). All cognitive aspects exist next to each other while being only valid in combination and common context. Thus, the analysis of construct relations and proven practical significance/predictive power is still purpose of research.

References

Andersen, P. B. (1997). *A theory of computer semiotics*. New York: Cambridge University Press.

Anderson, J. R. (2000). *Learning and memory: An integrated approach* (2nd ed.). New York: Wiley.

Atkinson, R. C., & Shiffrin, R. (1968). Human memory: A proposed system and its control processes. In K. W. Spence & J. T. Spence (Eds.), *The psychology of learning and motivation: Advances in research and theory* (Vol. 2). New York: Academic.

Atkinson, R. C., & Shiffrin, R. M. (1971). The control of short-term memory. *Scientific American, 225*, 82–90.

Baddeley, A. D. (1999). *Essentials of human memory*. Hove: Psychology Press.

Billings, C. E. (1997). *Aviation automation – The search for a human-centered approach*. Hillsdale: Lawrence Erlbaum.

Boff, K. R., & Lincoln, J. E. (1988). *Fatigue: Effect on performance. Engineering data compendium: Human perception and performance*. Wright Patterson AFB, Ohio: Harry G. Armstrong Aerospace Medical Research Laboratory.

Card, S. K., Moran, T. P., & Newell, A. (1983). *The psychology of human–computer-interaction*. Hillsdale: Lawrence Erlbaum.

Carroll, J. M., & Olson, J. R. (1988). Mental models in human–computer interaction. In M. Helander (Ed.), *Handbook of human computer interaction* (pp. 45–65). Amsterdam: Elsevier.

Dunckel, H. (1986). Handlungstheorie. In R. Rexilius & S. Grubitsch (Eds.), *Psychologie – Theorien, Methoden und Arbeitsfelder* (pp. 533–555). Hamburg: Rowohlt.

Endsley, M. R. (1988). Design and evaluation for situation awareness enhancement. *Proceedings of the human factors society 32nd annual meeting*, Human Factors Society, Santa Monica, pp. 97–101.

Endsley, M. R. (1995). Toward a theory of situation awareness. *Human Factors, 37*(1), 32–64.

Endsley, M. R. (2000). Theoretical underpinnings of situational awareness: A critical review. In M. R. Endsley & D. J. Garland (Eds.), *Situation awareness – Analysis and measurement*. Mahwah: Lawrence Erlbaum.

Fairclough, S. H., Tattersall, A. J., & Houston, K. (2006). Anxiety and performance in the British driving test. *Transportation Research Part F: Traffic Psychology and Behaviour, 9*, 43–52.

Flemisch, F. O. (2001). *Pointillistische Analyse der visuellen und nicht-visuellen Interaktionsressourcen am Beispiel Pilot-Assistenzsystem*. Aachen: Shaker Verlag.

Gaillard, A. W. (1993). Comparing the concepts of mental load and stress. *Ergonomics, 36*(9), 991–1005.

Gollwitzer, P. M. (1996). Das Rubikonmodell der Handlungsphasen. In J. Kuhl & H. Heckhausen (Eds.), *Motivation, volition und handlung* (Enzyklopädie der Psychologie: Themenbereich C, Theorie und Forschung: Ser. 4, Motivation und Emotion 4th ed., pp. 531–582). Bern/Göttingen/Seattle/Toronto: Hogrefe.

Grandjean, D., Sander, D., & Scherer, K. R. (2008). Conscious emotional experience emerges as a function of multilevel, appraisal-driven response synchronization. *Consciousness & Cognition, 17*(2), 484–495.

Hacker, W. (1978). *Allgemeine Arbeits- und Ingenieurpsychologie. Psychische Struktur und Regulation von Arbeitstätigkeiten*. Bern: Huber.

Hacker, W. (1986). *Arbeitspsychologie*. Berlin/Bern/Toronto/New York: Deutscher Verlag der Wissenschaften/Huber.

Hancock, P. A., & Szalma, J. L. (2008). Performance under stress. In P. A. Hancock & J. L. Szalma (Eds.), *Performance under stress*. Aldershot: Ashgate.

Hart, S. G., & Staveland, L. E. (1988). Development of a multi-dimensional workload rating scale: Results of empirical and theoretical research. In P. A. Hancock & N. Meshkati (Eds.), *Human mental workload* (pp. 139–183). Amsterdam: Elsevier.

Hart, S. G., & Wickens, C. D. (1990). Workload assessment and prediction. In H. R. Booher (Ed.), *Manprint. An approach to system integration* (pp. 257–296). New York: Van Nostrand Reinhold.

Heckhausen, H. (1989). *Motivation und Handeln*. Berlin: Springer.

Heckhausen, H., & Gollwitzer, P. M. (1987). Thought contents and cognitive functioning in motivational versus volitional states of mind. *Motivation and Emotion, 11*, 101–120.

Heckhausen, J., & Heckhausen, H. (2008). *Motivation and action*. New York: Cambridge University Press.

Herczeg, M. (1994). *Software-Ergonomie*. Bonn: Addison-Wesley.

Herczeg, M. (2000). Sicherheitskritische Mensch-Maschine-Systeme. *FOCUS MUL, 17*(1), 6–12.

Herczeg, M. (2002). Intention-based supervisory control – Kooperative Mensch-Maschine-Kommunikation in der Prozessführung. In M. Grandt & K.-P. Gärnter (Eds.), *Situation awareness in der Fahrzeug- und Prozessführung* (DGLR-Bericht 2002–04, pp. 29–42). Bonn: Deutsche Gesellschaft für Luft- und Raumfahrt.

Herczeg, M. (2004). Interaktions- und Kommunikationsversagen in Mensch-Maschine-Systemen als Analyse- und Modellierungskonzept zur Verbesserung sicherheitskritischer Technologien. In M. Grandt (Ed.), *Verlässlichkeit der Mensch-Maschine-Interaktion* (DGLR-Bericht 2004–03, pp. 73–86). Bonn: Deutsche Gesellschaft für Luft- und Raumfahrt.

Herczeg, M. (2006a). *Interaktionsdesign*. München: Oldenbourg.

Herczeg, M. (2006b). Differenzierung mentaler und konzeptueller Modelle und ihrer Abbildungen als Grundlage für das Cognitive Systems Engineering. In M. Grandt (Ed.), *Cognitive Engineering in der Fahrzeug- und Prozessführung* (DGLR-Bericht 2006–02, pp. 1–14). Bonn: Deutsche Gesellschaft für Luft- und Raumfahrt.

Herczeg, M. (2006c). Analyse und Gestaltung multimedialer interaktiver Systeme. In U. Konradt, & B. Zimolong (Hrsg.), Ingenieurpsychologie, Enzyklopädie der Psychologie, Serie III, Band 2, (S. 531–562). Göttingen: Hogrefe Verlag.

Herczeg, M. (2008). Vom Werkzeug zum Medium: Mensch-Maschine-Paradigmen in der Prozessführung. In M. Grandt & A. Bauch (Eds.), *Beiträge der Ergonomie zur Mensch-System-Integration* (DGLR-Bericht 2008–04, pp. 1–11). Bonn: Deutsche Gesellschaft für Luft- und Raumfahrt.

Herczeg, M. (2009). *Software-Ergonomie. Theorien, Modelle und Kriterien für gebrauchstaugliche interaktive Computersysteme* (3rd ed.). München: Oldenbourg.
Herczeg, M. (2010). Die Rückkehr des Analogen: Interaktive Medien in der Digitalen Prozessführung. In M. Grandt & A. Bauch (Eds.), *Innovative Interaktionstechnologien für Mensch-Maschine-Schnittstellen* (pp. 13–28). Bonn: Deutsche Gesellschaft für Luft- und Raumfahrt.
Hockey, G. R. (1986). Changes in operator efficiency as a function of environmental stress, fatigue and circadian rhythms. In K. R. Boff, L. Kaufmann, & J. P. Thomas (Eds.), *Handbook of perception and human performance* (Cognitive processes and performance, Vol. II). New York: Wiley.
Immelmann, K., Scherer, K. R., & Vogel, C. (1988). *Psychobiologie. Grundlagen des Verhaltens.* Weinheim: Beltz-Verlag.
Johnson, A., & Proctor, R. W. (2003). *Attention: Theory and practice*. London: Sage.
Kahneman, D. (1973). *Attention and effort*. Englewood Cliffs: Prentice Hall.
Lazarus, R. S. (1974). Psychological stress and coping in adaption and illness. *International Journal of Psychiatry in Medicine, 5*, 321–333.
Lazarus, R., & Folkman, S. (1984). *Stress, appraisal, and coping*. New York: Springer.
Miller, G. A. (1956). The magical number seven, plus or minus two: Some limits on our capacity for processing information. *Psychological Review, 63*(2), 81–97.
Miller, G. A., Galanter, E., & Pribram, K. H. (1960). *Plans and the structure of behavior*. New York: Holt, Rinehart & Winston.
Nake, F. (2001). Das algorithmische Zeichen. In W. Bauknecht, W. Brauer, & T. Mück (Eds.). *Tagungsband der GI/OCG Jahrestagung 2001*, Bd. II, (pp. 736–742), Universität Wien.
NATO. (2001). Research and technology organisation. The Human Factors and Medicine Panel (HFM) Research and Study Group 24. *RTO-TR-021 NATO Guidelines on Human Engineering Testing and Evaluation*. WWW-Document. http://www.rto.nato.int/panel.asp?panel=HFM&O=TITLE&topic=pubs&pg=3#
Navon, D., & Gopher, D. (1979). On the economy of the human information processing system. *Psychological Review, 86*, 214–255.
Newell, A., & Simon, H. (1972). *Human problem solving*. Englewood Cliffs: Prentice Hall.
Nordwall, B. D. (1995). Military cockpits keep autopilot interface simple. *Aviation Week & Space Technology, 142*(6), 54–55.
Norman, D. A. (1986). Cognitive engineering. In D. A. Norman & S. W. Draper (Eds.), *User centered system design* (pp. 31–61). Hillsdale: Lawrence Erlbaum.
Norman, D., & Bobrow, D. (1975). On data-limited and resource-limited processing. *Journal of Cognitive Psychology, 7*, 44–60.
Nöth, W. (2000). *Handbuch der Semiotik*. Stuttgart-Weimar: Metzler.
Pinel, J. (1993). Biopsychology (2nd ed.). Allyn and Bacon, Boston.
Rasmussen, J. (1983). Skills, rules, and knowledge; signals, signs, and symbols, and other distinctions in human performance models. *IEEE Transactions on Systems, Man, and Cybernetics, SMC-13*(3), 257–266.
Rasmussen, J. (1984). Strategies for state identification and diagnosis in supervisory control tasks, and design of computer-based support systems. In W. B. Rouse (Ed.), *Advances in man-machine systems research* (Vol. 1, pp. 139–193). Greenwich: JAI Press.
Rasmussen, J. (1985a). The role of hierarchical knowledge representation in decision making and system management. *IEEE Transactions on Systems, Man, and Cybernetics, SMC-15*(2), 234–243.
Rasmussen, J. (1985b). *A framework for cognitive task analysis in systems design*. Report M-2519, Risø National Laboratory.
Rasmussen, J. (1986). *Information processing and human–machine interaction*. New York: North-Holland.
Rasmussen, J., Pejtersen, A. M., & Goodstein, L. P. (1994). *Cognitive systems engineering*. New York: Wiley.

Sanders, A. F. (1983). Towards a model of stress and human performance. *Acta Psychologica, 53*, 61–67.
Scheffer, D., & Heckhausen, H. (2008). Trait theories of motivation. In J. Heckhausen & H. Heckhausen (Eds.), *Motivation and action*. New York: Cambridge University Press.
Scherer, K. R. (1984). On the nature and function of emotion: A component process approach. In K. R. Scherer & P. Ekman (Eds.), *Approaches to emotion* (pp. 293–317). Hillsdale: Erlbaum.
Scherer, K. R. (1990). Theorien und aktuelle Probleme der Emotionspsychologie. In K. R. Scherer (Ed.), *Enzyklopädie der Psychologie (C, IV, 3)* (Psychologie der Emotion, pp. 1–38). Göttingen: Hogrefe.
Schneider, K., & Dittrich, W. (1990). Evolution und Funktion von Emotionen. In K. R. Scherer (Ed.), *Enzyklopädie der Psychologie, Teilband C/IV/3* (Psychologie der Emotion, pp. 41–114). Göttingen: Hogrefe.
Schneider, K., & Schmalt, H.-D. (2000). *Motivation*. Stuttgart: Kohlhammer.
Seitz, D. P., & Häcker, H. O. (2008). Qualitative befundanalyse des Konstruktes situational awareness – A qualitative analysis. In Bundesministerium der Verteidigung – PSZ III 6 (Ed.), *Untersuchungen des Psychologischen Dienstes der Bundeswehr* (Jahrgang, Vol. 43). Bonn: Verteidigungsministerium.
Sheridan, T. B. (1987). Supervisory control. In G. Salvendy (Ed.), *Handbook of human factors* (pp. 1243–1268). New York: Wiley.
Sheridan, T. B. (1988). Task allocation and supervisory control. In M. Helander (Ed.), *Handbook of human–computer interaction* (pp. 159–173). Amsterdam: Elsevier Science Publishers B.V. (North Holland).
Simon, H. (1981). *The sciences of the artificial*. Cambridge: MIT Press.
Stein, M. (2002). Entwicklung eines Modells zur Beschreibung des Transferprozesses am Beispiel Arbeitsschutz. In Bundesanstalt für Arbeitsschutz und Arbeitsmedizin (Ed.), *Schriftenreihe der Bundesanstalt für Arbeitsschutz und Arbeitsmedizin*. Dortmund/Berlin: Wirtschafts-Verlag.
Stein, M. (2008). Informationsergonomie. Ergonomische Analyse, Bewertung und Gestaltung von Informationssystemen. In Bundesanstalt für Arbeitsschutz und Arbeitsmedizin (Hrsg.), Dormund: Druck der Bundesanstalt für Arbeitsschutz und Arbeitsmedizin.
Stein, M., & Müller, B. H. (2005). Motivation, Emotion und Lernen bei der Nutzung von internetbasierten Informationssystemen. In Gesellschaft für Arbeitswissenschaft (Ed.), *Personalmanagement und Arbeitsgestaltung* (Bericht zum 51. Kongress der Gesellschaft für Arbeitswissenschaft vom 22.–24. März 2005, pp. 665–668). Dortmund: GfA Press.
Stein, M., Müller, B. H., & Seiler, K. (2002). Motivationales Erleben bei der Nutzung von internetbasierten Informationssystemen. In R. Trimpop, B. Zimolong, & A. Kalveram (Eds.), *Psychologie der Arbeitssicherheit und Gesundheit – Neue Welten – Alte Welten 11. Workshop "Psychologie der Arbeitssicherheit", Nümbrecht, 2001* (p. 330). Heidelberg: Roland Asanger.
Szalma, J. L., & Hancock, P. A. (2002). *Mental resources and performance under stress*. WWW Dokument Verfügbar unter: http://www.mit.ucf.edu/WhitePapers/Resource%20white%20paper.doc.
Treisman, A. M. (1969). Strategies and models of selective attention. *Psychological Review, 76*, 282–292.
Tsang, P.S. & Vidulich, M.A. (2003). Principles and Practice of Aviation Psychology. Mahwah, NJ: Lawrence Erlbaum Association.
Vicente, K. J. (1999). *Cognitive work analysis*. Hillsdale: Lawrence Erlbaum.
Volpert, W. (1974). *Handlungsstrukturanalyse als Beitrag zur Qualifikationsforschung*. Köln: Pahl-Rugenstein.
von Cranach, M., & Kalbermatten, U. (1982). Zielgerichtetes Alltagshandeln in der sozialen Interaktion. In W. Hacker et al. (Eds.), *Kognitive und motivationale Aspekte der Handlung* (pp. 59–75). Bern: Huber.
Waag, W. L., & Bell, H. H. (1997). Situation assessment and decision making in skilled fighter pilots. In C. E. Zsambok & G. Klein (Eds.), *Naturalistic decision making* (pp. 247–254). Mahwah: Lawrence Erlbaum.

Ware, C. (2000). *Information visualization – Perception for design*. San Diego: Academic.
Wickens, C. D. (1980). The structure of attentional resources. In R. Nickerson (Ed.), *Attention and performance VIII* (pp. 239–257). Hillsdale: Erlbaum.
Wickens, C. D. (1984). Processing resources in attention. In R. Parasuraman & D. R. Davies (Eds.), *Varieties of attention* (pp. 63–102). New York: Academic.
Wickens, C. D. (2008). Multiple resources and mental workload. *Human Factors, 50*, 449–455.
Wickens, C. D., & Hollands, J. G. (2000). *Engineering psychology and human performance*. Upper Saddle River: Prentice Hall.

Automotive

Klaus Bengler, Heiner Bubb, Ingo Totzke, Josef Schumann, and Frank Flemisch

1 Introduction

The main focus in this chapter shall be on information and communication in the situational context "vehicle". Thereby, based on relevant theoretical approaches in the area of "information" and "communication", it shall be demonstrated that the topic communication and information becomes even more complex through the special situation in a car while driving. First, one needs to distinguish between "natural" human–human communication in a vehicle (e.g. passenger, mobile telephony) and the more "artificial" human–vehicle communication. For designing the optimum human–vehicle communication, findings and approaches of the human–human communication thus shall be used. Information and communication between drivers and the impact on the traffic system shall not be taken into consideration here.

This paper will expressly underline that communication and information in a vehicle includes more than using a mobile phone or the internet while driving and the question of distraction of the driver thus arising. So, first of all, it becomes obvious that, in a vehicle, various sources of information and communicating partners are relevant in order to successfully manage the task of driving: from road construction information (e.g. signage, traffic light regulation) via information coming from other road users within the direct surrounding of a vehicle

K. Bengler (✉) • H. Bubb
Lehrstuhl für Ergonomie, Technische Universität München, Munich, Germany

I. Totzke
Lehrstuhl für Psychologie III, Interdisziplinäres Zentrum für Verkehrswissenschaften an der Universität Würzburg, Würzburg, Germany

J. Schumann
Projekte ConnectedDrive, BMW Group Forschung und Technik GmbH, Munich, Germany

F. Flemisch
Institut für Arbeitswissenschaft, RWTH Aachen, Aachen, Germany

M. Stein and P. Sandl (eds.), *Information Ergonomics*, DOI 10.1007/978-3-642-25841-1_4,

(e.g. eye contact or hand signals to settle traffic in case of unclear traffic situations), to information from the vehicle to its driver, e.g. by advanced driver assistance systems or driver information systems and warning messages. Thus, the driver is confronted with different communication situations and, in particular, communication contents.

Moreover, apart from different variants of spoken language, sophisticated sign and symbol systems play an important role concerning the information and communication of the driver. Besides road construction signs and symbols (e.g. traffic signs or routing and configuration of streets and roads), nonverbal signs of passengers or other drivers need to be considered too. After all, one should not forget that in the context of driving it is not possible either *not* to communicate: "Activity or inactivity, words or silence: all have message value" (Watzlawick et al. 1996, p. 51). Therefore, even a lack of explicit information or acts of communicating is a source of information for the driver. That means that he/she knows for example that certain traffic rules apply even when there are no signs along the route, or that missing information from the vehicle indicate its functional efficiency. It is also other road users' behavior or the lack of a particular behavior that provides important information in order to continue driving.

Furthermore, technological progress has a significant impact on information and communication inside the vehicle. First of all, the development of radio technology massively influenced the communication situation in the vehicle. Through this development, a larger group of drivers could be provided continuously with most varied information via radio. It took more than 50 years, however, from the first car radio (1922) to the traffic decoder (1973), a radio that analyzes station identifications and road messages and is thus able to provide the driver with precise traffic information. Those asynchronous, unidirectional communication sources (e.g. broadcast radio) were expanded by synchronous, bidirectional communication sources (e.g. telephony, wireless radio). Thus, the recipient of a message has the possibility to transmit a message to the transmitter. The development of CB radio systems and GSM mobile telephony later on allowed for bidirectional communication.

In-vehicle information and communication do indeed present a special case concerning general scientific information and communication models, but are certainly not a general exception. Thus, it makes sense to look at information and communication in the vehicle context with the corresponding theoretical models and terms. As there are several terms that by now have entered the ISO/DIN standards, the corresponding approaches – if any – shall be used. As a summary, possibilities and limits of the applying theoretical approaches shall be presented (see Sect. 2.4).

2 A Working Definition of Information and Communication

The term "communication" is basically very ambiguous. Merten (1993) for example refers to more than 160 definitions that all deal with the term "communication". According to Maletzke (1998), today one uses a concept of the term "communication" that is based on the idea of a reciprocal comprehension process between partners with a common basis. Consequently, it is assumed that there must be a common stock of symbols (shared code) between transmitter and recipient as a minimal communication requirement.

While such a definition specifically refers to human communication, other definitions expand the term "communication", e.g. to the field of human–machine communication. Corresponding statements on definitions lead to the specification of the term "communication" in the form of ISO standards for the vehicle context. ISO standard EN ISO 15005 (2002) for example defines communication very generally as an exchange or a transmission of information, or a dialog as an exchange of information between driver and system in order to achieve a certain objective, which is initiated either by the driver or the system and consists of a sequence of interlinked operation actions which can comprise more than one modality. Simplified, information thus is seen as elements or units of communication, which need to be exchanged between communicating partners.

2.1 Communication Models

By now, there is a whole range of models dealing with communication (to get an overview, cf. Krallmann and Ziemann 2001; Maletzke 1998; Steinbuch 1977). Most approaches refer to Lasswell's formula (1964): "Who says what in which channel, to whom, with what effect?" (p. 38). Communication thus consists of a total of five basic units: communicator ("Who?"), information ("What?"), medium ("Which channel?"), recipient ("To whom?") and effect ("What effect?"). Critics say in return that the effects of communication have to be seen as one aspect of the recipient and thus do not form an independent basic unit of communication. Consequently, this resulted in four basic units forming communication (Maletzke 1998).

Closely related to the so called Lasswell Formula is the mathematical communication theory of Shannon and Weaver (1949, see Fig. 1).

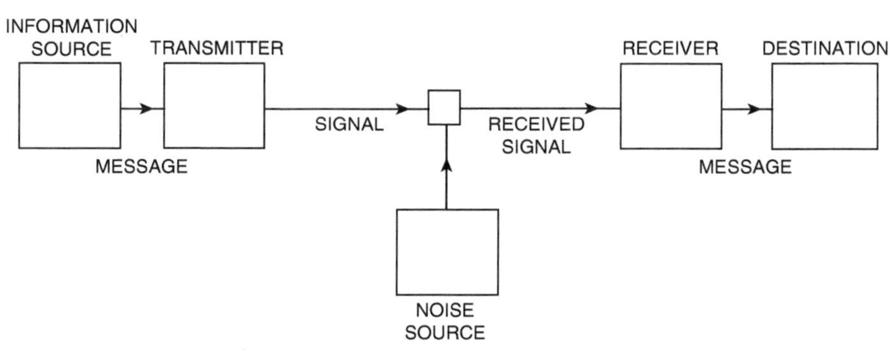

Fig. 1 Communication model according to Shannon and Weaver (1949)

According to this, a message producer (transmitter) has a meaning supply, a symbol supply, and rules for matching the symbol elements the meaning elements (“code”). It wants to transmit a sequence of meaning elements (“message”). In order to do so, it translates the message into a sequence of symbols according to the matching rules (“encoding”). The symbol sequence is being transmitted as a signal via a channel, which in turn is liable to interferences. The recipient eventually decodes the signal by translating the received symbol sequence, using its matching rules, into the corresponding meaning sequence (“decoding”). Following this model, communication is successful if transmitter and recipient (roughly) have the same supply of meanings, symbols and matching rules and if the signal is not being destroyed on its way. This communication model assumes a unidirectional linear transmitting process as a kind of “one-way street”.

Communication, however, always also comprises the exchange of information, i.e. the recipient’s reaction, which then again is received by the original transmitter, etc. According to the so-called organon model by Buehler (1934), it is only through these reciprocal processes between transmitter and recipient that meaning arises within communication. If we followed this model, we would have to describe the communication situation through transmitter, recipient and topic. However, it becomes obvious soon that this model describes the act of communicating very comprehensively but at the same time very generally and rarely differentiated.

It is furthermore assumed that communication is always target-oriented. For language-based communication, the assumption arises that if a person wants something, he/she can achieve it through speaking and thus makes a certain comment. Thus, it would not be single words or sentences that form the fundamental units of communication, but holistic acts of speech. According to Searle (1982), one has to distinguish between five general forms of language use:

- Representative (e.g. to claim, determine, report)
- Directives (e.g. to command, demand, ask)
- Commissives (e.g. to promise, guarantee, vow)
- Expressives (e.g. to thank, congratulate, condole)
- Declarations (e.g. to appoint, nominate, convict)

Grice (1967/1989, 1975/1993) looks at human communication even more holistically, concentrating on the speaker’s intention and the intended response of the listener. Grice thus sees communication as a holistic process. Therefore, the requirement is that for a clarification of linguistic term usage and the specific character of communication, it is necessary to always examine them in different usage contexts (Krallmann and Ziemann 2001). As a maxim for successful communication, Grice introduces the following:

- *Maxim of Quantity*: Make your contribution as informative as is required for the current purposes of the exchange. Do not make your contribution more informative than is required.
- *Maxim of Quality*: Do not say what you believe to be false. Do not say that for which you lack adequate evidence.

- *Maxim of Relation*: Be relevant.
- *Maxim of Manner*: Avoid obscurity of expression. Avoid ambiguity. Be brief. Be orderly.

This model postulates at the same time that communicating partners usually follow these maxims for cooperative and successful communication, or expect their corresponding partners to follow those maxims.

2.2 Evaluation of the Communication Models

For a comprehensive examination of the effects of communication, one thus has to consider various effect aspects. Steinbuch (1977) for example lists the following effect aspects:

1. *Physical-physiological-technical aspects*: how are the signals that are needed for the transportation of information transferred from transmitter to recipient?
2. *Psychological aspects*: under which circumstances will the recipient perceive and process the signals?
3. *Semantic aspects*: which correlations have to exist between information and signal in order to have the recipient extract the adequate information from the signals?

These effect aspects are widely covered by the prevailing communication models. It becomes obvious, however, that these models describe the act of communication very comprehensively but mostly still very generally. Thus, one cannot just simply transfer them to the vehicle context and human–human, or human–vehicle communication there. For example, they do not account for time aspects or the interruptibility of the communication act, if necessary, which plays an extremely important role to avoid distraction especially while driving (Praxenthaler and Bengler 2002). In contrast to other communication situations, the relevancy and validity of some pieces of information in the vehicle context depend on reception time and cruising speed. A navigation signal, for example, needs to be spoken at a certain moment before turning, if the driver is supposed to be able to make use of it. Road signs, too, need to be perceived, interpreted, and implemented as the information content otherwise becomes obsolete. It is especially this temporal dynamics that is mainly prevalent in mobile communication situations and depends strongly on the temporal characteristics of the traffic situation. Accordingly, the fourth aspect for examining communication in the vehicle context arises.

4. *Temporal aspects*: when does the transmitter have to emit information so that the recipient can react adequately and without time pressure at the right point in time?

In the following, these effect aspects of in-vehicle communication shall be considered in particular and in detail.

3 Information and Communication Against the Background of the Driving Task

Information and communication in the vehicle are generally seen as an additional burden to the driver that, depending on his/her skills and abilities, lead to a resulting load. This definition is based on the so-called *stress–strain* concept.

Up to now, the most important evaluation criterion in a complex human–machine system has been the *workload* of the operator with respect to his/her information processing capabilities. Workload is used to describe the cost of accomplishing task requirements for the human element of human–machine systems (Hart and Wickens 1990).

Through the introduction of automation in the cockpit the role of the pilot changed, however, from an actor to a supervisor of (partly) automated system conditions. In such an automated system, the pilot operates in a specific task environment and sets initial conditions for intermittently adjusting and receiving information from a machine that itself closes the control loop through external sensors and effectors. This new quality of control was termed "*supervisory control*" (Sheridan 1997) and emphasized the need of pilot awareness of the complex and dynamic events which take place during flight, and how the pilot could use this processed information for possible future flight control actions. In aviation, the notion of "*situation awareness*" evolved to denote the degree of adaptation between a pilot and the cockpit environment (Endsley 1995; Flach and Rasmussen 2000). Situation awareness describes the pilot's awareness of the present conditions in a human–machine system, i.e. situation and system states, and the ability of the pilot to integrate this knowledge into adequate pilot control actions.

As stated in the introduction, in human factors the term workload has evolved to describe the influence of the task load on the operator in a human–machine system. The workload of the operator is dependent on the task load (i.e. task requirement) and the corresponding abilities of the operator (e.g. skills, state). The workload retroacts on the operator and changes or modifies the abilities of the operator (e.g. fatigue, learning). This mutual dependency has been expressed in the stress–strain concept (German: "Belastungs-Beanspruchungskonzept"; see Rohmert and Breuer 1993; Kirchner 2001).

The operator acts in a specific environment and tries to accomplish the task requirements. The operator is able to monitor the performance of the human–machine system, which is indicated through the feedback loop in Fig. 3.1. Three different sources of load can be identified:

- Environmental load or stressors (e.g. illumination, noise, climate),
- Task load (e.g. time pressure, task set switches, cognitive task level),
- Machine load (e.g. information from different interface sources).

All these load sources result in workload depending on the operator's individual attributes (state and expertise) and coping strategies (see also the technical standard on cognitive workload; DIN EN ISO 10075 2000).

At the same time, it is assumed that human processing capacities are limited, i.e. it is not possible to perceive and process an unlimited amount of information simultaneously, or to respond adequately to the information. Accordingly, the necessity arises to optimize information and communication in the vehicle context so that only a minimum of negative effects (strain) on driving and road safety occur when driving a vehicle. It is quite obvious that it is not only redundant information that results in stress, but that there is also necessary information that has to be there in order to accomplish the driving task. I.e. missing information can also lead to strain as drivers who do not know their whereabouts, for instance, have to put much more energy into orienting themselves without navigation information from the vehicle. Thus, corresponding theoretical models shall be briefly described to explain possible interferences between the driving task as well as information and communication. It can be expected that these mutual influences depend on features of cognitive processes that are involved in information processing. In conclusion, the diversity of the communication context "vehicle" shall be illustrated by means of an in-vehicle information and communication taxonomy.

The theoretical background for the explanation of possible attention deficits on the driver's side and the consequent losses in the driving ability is mostly capacity and resource models that hold the assumption of one person having only limited capacities to process information. If those available information processing capacities are exceeded, the acting person's performance will deteriorate. Currently, assumptions of the multiple resources model, enhanced by a so called bottleneck on the processing level of the reaction selection (i.e. the transfer of encoded stimuli into precise action orders) are to be supported. This results in the assumption that the selection of various reaction alternatives is attention-dependent.

Particularly strong interferences between driving task and information/communication are to be expected in case of attention-related control processes on the knowledge level.

While in case of non-familiar events it is mainly these attention-related control processes that are activated, it is rather scheme-oriented control processes that are activated as the events become more familiar, firstly on the rule and habit level. As a result, the respective information turns more and more into a signal which leads to faster information processing but increasing error-proneness. At the same time, one should consider that the driver is exposed to numerous different sources of information and communication partners while driving which interact with the driver on different processing channels and with different speech ratios.

In the vehicle, the driver's attention and its optimum allocation to aspects relevant to action in the context of the driving situation play a significantly more important role than in other environments and situations. For example, lack of attention to the vehicle in front entails the danger of colliding with this vehicle if it suddenly would brake. After all, more complex attention and information processing processes are necessary on a highway, for example to observe vehicles in front to assess whether sudden traffic jams with major variations in driving speed could arise. Numerous studies prove that, especially under information and communication conditions, the driver's alertness is significantly diverted from the driving task towards information and communication (e.g. Atchley and Dressel 2004; Byblow 1990; Mutschler et al. 2000; Sperandio and Dessaigne 1988; Verwey 1993; Verwey and Veltman 1995; Vollrath and Totzke 2000).

The theoretical background to explain possible attention deficits of the driver and the consequent reduction of driving capabilities are mostly capacity models (e.g. Broadbent 1958; Deutsch and Deutsch 1963; Keele 1973; Welford 1952) or resource models (e.g. Kahneman 1973; Moray 1967; Wickens 1980, 1984, 2002).

Within all of these models one can find the assumption that a person has only a limited capacity to process information. If these available capacities for information processing are exceeded, the consequent mental overload leads to a decrease in the acting person's performance. Therefore, one can expect reduced driving capabilities through information and communication in the vehicle, if the requirements by the information and communication exceed the driver's limit of information processing.

Depending on the theoretical models, these information processing capacities are described e.g. as general processing capacity that can be allotted to a task (Kahneman 1973; Moray 1967), or as multiple resources that a person has available and he/she can allocate to one or several tasks (simultaneously) (Wickens 1980, 1984, 2002). Accordingly, predictions can be made as to when and in which form in-vehicle information and communication is recommendable:

- Models of a general processing capacity suggest that information and communication can be processed particularly sequentially (i.e. alternately) with special tasks while driving (e.g. braking, accelerating, steering) as those are increasingly accomplished automatically by the experienced driver. Thus, information and communication should only be allowed if no other processing acts relevant for driving are necessary.
- Models of multiple resources, however, allow for the conclusion that requirements on the driver's information processing which arise from information and communication influence the interference between driving and information/communication significantly. Thus, especially acoustic information should be provided in the vehicle that requires, if at all, a vocal response as they compete only minimally with the driving task which is mainly controlled visually.

By now, there are numerous empirical findings that support the model of multiple resources (to get an overview see e.g. Wickens 1980, 1984, 2002; Wickens and Hollands 2000). On the other hand, it has been shown that a perfect time sharing strategy between tasks that have to be accomplished simultaneously cannot be achieved. Braune and Wickens (1986) e.g. prove that high requirements on a person through two different tasks can cause strong interferences, even though no common resources are needed according to the multiple resources model. This led to articulating the assumption that attention is reduced to a one-channel system in case of high requirements. This model thus maybe has to be extended by one aspect of capacity models: the so called "bottleneck". It is assumed that such a "bottleneck" occurs on the processing level of reaction selection (i.e. the transfer of encoded stimuli to precise action orders) (e.g. Pashler 1994; Welford 1967). This results in a dependency on attention for selecting different reaction alternatives so that only one stimulus can be processed at a time. Perception, encoding and intermediate storage of various stimuli on the other hand occurs simultaneously in the working memory. By now, there are numerous findings supporting this assumption (Meyer and Kieras 1997).

According to Rasmussen (1983), attention-related information processing can be expected particularly in case of knowledge-based control processes. Such processes comprise analyses of original and target state, planning and decision, as well as selection and usage of means to reach the goal and check the applied strategy. In the vehicle context, that comprises the stabilization of the vehicle by a student driver, driving on slippery or icy roads, and the driver's orientation in a foreign city (Schmidtke 1993). Successfully repeating an action leads to shifting the control processes to a subordinate level. Thus, more scheme-oriented control processes occur with an increasing degree of familiarity, first on the so called rule level. Processing then can be described as if-then rules (Norman 1983). In the vehicle context, that comprises overtaking one or several vehicles, as well as choosing between alternative routes which are already known to the driver (Schmidtke 1993). Eventually, so called rule-based control processes become increasingly active where pre-programmed action schemes are retrieved through the corresponding signals from the environment. Signals do not have to be interpreted but have an immediate trigger function for action schemes (Norman 1983). In the area of driving, there are the following examples: accurate driving through bends or turning at familiar junctions as well as driving the daily trip to work (Schmidtke 1993). With increasing familiarity, information gets more and more the character of signals, which means that the speed of information processing increases while the susceptibility to failure in case of exceptions from the rules, however, also increases. The examples taken from the vehicle context highlight the necessity of distinguishing between attention-related and scheme-oriented behavior as well as possible deficits in in-vehicle communication situations.

3.1 Taxonomy of Information and Communication

As already described, the driver is exposed to a variety of different sources of information or communication partners while driving which interact with the driver via different processing channels or speech ratios. Table 1 tries to display this complexity of in-vehicle information and communication possibilities with the aid of taxonomy.

This taxonomy demonstrates that the driver is provided with information through external sources (e.g. road signs, car radio, DAB, CD), other people (e.g. mobile phone, passenger, wireless, other road users), as well as the vehicle itself (e.g. failure indication, instrument display, labeling, speech recognition systems, navigation advice). Information through the vehicle that is useful for

Table 1 Taxonomy of information and communication

	Transmitter	Channel	Local closeness to event	Temporal closeness to event	Encoding
External sources					
Signage	Sign, poster, traffic lights, changing road signs	Visual	High	Asynchronous	Symbols, written signs
Car radio	PBS	Acoustic	Low	Synchronous	Natural speech
DAB	PBS	Acoustic, visual	Low	Synchronous	Natural speech
Human/human					
Telephony	Human, machine	Acoustic	Low	Synchronous	Natural speech
Passenger	Human	Acoustic, visual	High	Synchronous	Natural speech
Wireless	Human	Acoustic	Low	Synchronous	Regulated natural speech
Communication between road users	Human	Visual		acoustic	High
Synchronous	Turn/horn signal, gesture, mimic				
Human/vehicle					
Error indication	Machine	Acoustic, visual	High	Synchronous	Alert signs writing symbols
Instrumentation	Machine	Acoustic, visual, haptic	High	Synchronous	Alert signs writing symbols
Labeling	Machine	Visual	High	Synchronous	Writing symbols
Speech dialog system	Machine	Acoustic, visual	High	Synchronous	Regulated natural speech, natural speech
Navigation	Machine	Acoustic, visual	High	Synchronous	Regulated natural speech, writing symbols

fulfilling the driving task can be e.g. driving-related information about the route or the vehicle's condition that are mainly transmitted via visual signals and symbols in a static or dynamic way. Acoustic signals are mainly used for alarms or advice (e.g. by the navigation system).

Apart from the heterogeneity of possible information and communication sources as well as the corresponding channels, it becomes obvious that sources, contents and time of communication are only partly synchronized. So basically, the driver can be simultaneously exposed to information through road signs (e.g. speed limit), the car radio (e.g. traffic jam report), mobile phone (e.g. call of a business partner), and the vehicle (e.g. failure indication "low oil level"). The requirements on the driver as a communication partner vary accordingly. These interruptions can only partly be solved technically through information management and priority rules. In case of such priority rules, principles have to be formulated for output logics (e.g. traffic safety as the main goal for information display with minimum distraction effects), as well as information on the output order (e.g. safety-critical information more important than comfort-oriented information), and approaches to avoid "information jams" (e.g. restriction to only little but useful information) (Färber and Färber 1999). One distinct example in this context is the mute function for audio sources in case of traffic or navigation announcements. In the remaining cases, prioritization lies within the driver's responsibility and can lead to errors. According to Norman (1981, 1986) and Reason (1979, 1986), in this case, mistakes in accomplishing the task can occur (e.g. omitting steps and information).

3.2 Vehicle Systems

The introduced taxonomy of in-vehicle information and communication (see Table 1) shows that human–vehicle communication takes up a significant part and is necessary to successfully accomplish the driving task. The vehicle provides the driver with visual, auditive, and particularly haptic and kinesthetic information. The following part, however, mainly concentrates on language and visual information and communication.

As the following chapter will first of all deal with general design principles for human–vehicle communication, this chapter shall concentrate on possibilities for optimization.

For further information concerning optimizing the physical aspects of human–vehicle communication see the corresponding design principles, or guidelines, or ISO/DIN standards.

4 Advanced Driver Assistance Systems and Haptic Controls

Information and communication pose a clear challenge regarding advanced driver assistance systems which support the driver in accomplishing the driving task by means of headway control systems (ACC, Distronic, etc.) or lane keeping assistants. In order to make use of these driver assistance systems, which have grown considerably in number and complexity, corresponding interaction concepts are needed. It is obviously not enough anymore to just have 'some' buttons. While some control systems like ABS or TCS function without sophisticated control elements, it is much more complex already to operate the cruise control. Future vehicles will increasingly be assisted by driver assistance systems in addition to control through the driver and will be moved highly automated, which calls for an appropriate design of human–vehicle communication. Automobile manufacturers have integrated assistance systems for quite some time now. These assistance systems do not only enhance comfort and manageability but intervene actively into the driving task and thus into the driver-vehicle system. Some systems, like e.g. ACC (Adaptive Cruise Control) or LKAS (Lane Keeping Assistant System), are already able to accomplish single parts of the main driving task autonomously. Similar to aviation, the degree of automation increases continually (Hakuli et al. 2009; Winner et al. 2006; Flemisch et al. 2003, 2008).

For the purpose of an ergonomic driver-vehicle communication, it is thus decisive that information and communication ensure the driver to still remain informed sufficiently about the system status and driving process within the driving loop. In order to guarantee this, it is useful to apply active controls.

In general terms, these controls are characterized by their nature of being able to generate forces themselves. To do so, they have active elements, like e.g. electric motors. The user can feel the thus generated forces on the control. Through appropriate activation, it is hence possible to provide the driver with information relevant for the task accomplishment (Bolte 1991). Looking at conventional driving without assistance systems first, an example would be being informed about the current speed when accelerating, not about the position of the throttle valve. These vehicle status quantities can additionally be superimposed with features assisting haptic communication. For technical purposes, two concepts of active control elements must be generally distinguished: force and position reflective control elements (Penka 2000). With force reflective control elements, the path or the angle of the control elements indicated by the user represent the setpoint setting for the system that has to be adjusted. Feedback information is returned by the force that corresponds to the system status. For example, by applying a deflection proportional force with mass and damping forces a spring characteristic can be simulated. The advantage of this principle lies in the variable characteristic's design of this mass-spring damper system. Hence, the operator can be provided with continuous feedback information about system status and system dynamics, dependent on the behavior of the control system (Fig. 2).

The application of unconventional active control elements in modern vehicles is possible due to the highly developed technology, especially in the field of by-wire systems, where the mechanical coupling of control element and object to be controlled is replaced by an electronic connection (Fig. 3).

Uncoupling the control element from the mechanical framework involves a number of advantages.

A complete redesign of the interface between driver and vehicle is possible, which is not based on technical restrictions anymore, but much rather on ergonomic design maxims. Number, geometry and arrangement of the control elements as well as their parameterization can be optimized regarding anthropometry and system ergonomics. In order to guarantee human machine communication to be as intuitive as possible, the dimensionality of the control element should be in accordance with the dimensionality of the task from an ergonomic point of view. A two-dimensional control, e.g. a side stick, should thus be attributed to the two-dimensional task of driving. Comparative studies (Eckstein 2001; Kienle et al. 2009) show that, with

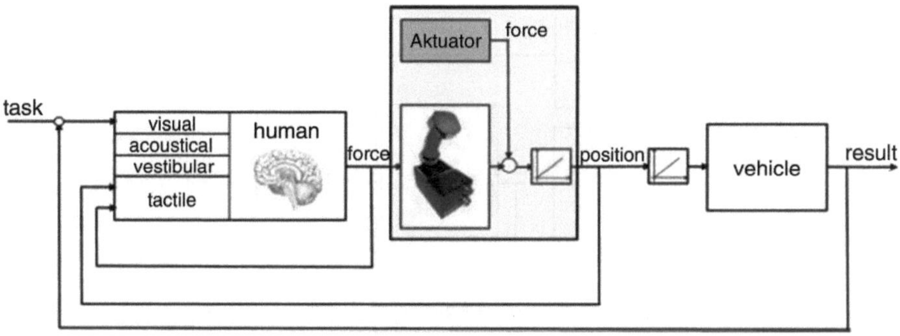

Fig. 2 Force reflective, active control elements (Source: Kienle et al. 2009)

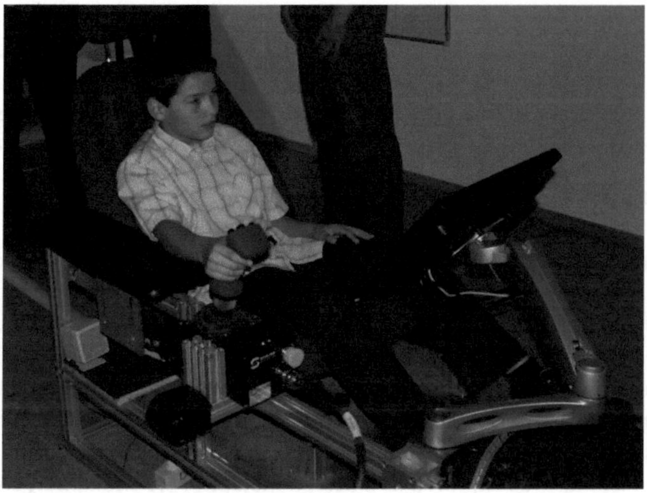

Fig. 3 H-Mode setup with active control (*side stick*)

adequate parameterization, a side stick helps to fulfill the driving task at least in a similar way like with conventional control elements, and in some cases even better.

While by-wire systems have a couple of advantages regarding the degrees of freedom, haptically active controls are still possible even with conventional cars mechanical connection between the steering wheel and the wheels: Active acceleration pedals are already in serial development (e.g. by Continental), and in most cars the steering wheel can be actively influenced or often steered with the electric power steering (Fig. 4).

Haptically active interfaces can be quite beneficial in conjunction with automation: Here the automation applies additional forces or torques to the interface. This force is added to the force of the human, leading to a shared control between the human and the automation (e.g. Griffith and Gillespie 2005; Mulder et al. 2011). If the automation is more sensible to the driver than just adding forces e.g. by having a

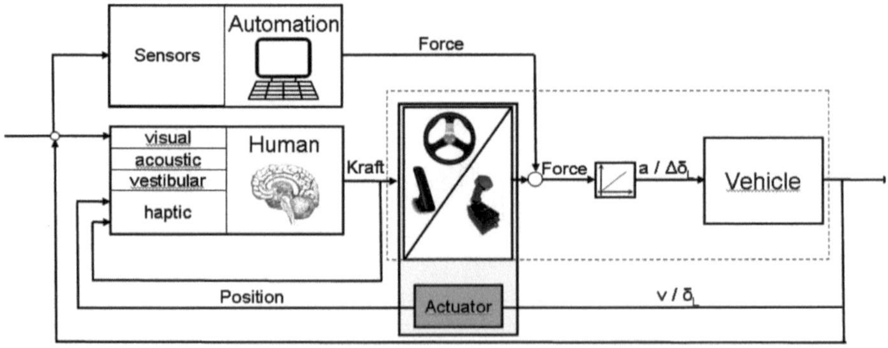

Fig. 4 Cooperative guidance and control with haptic interfaces (Kienle et al. 2009)

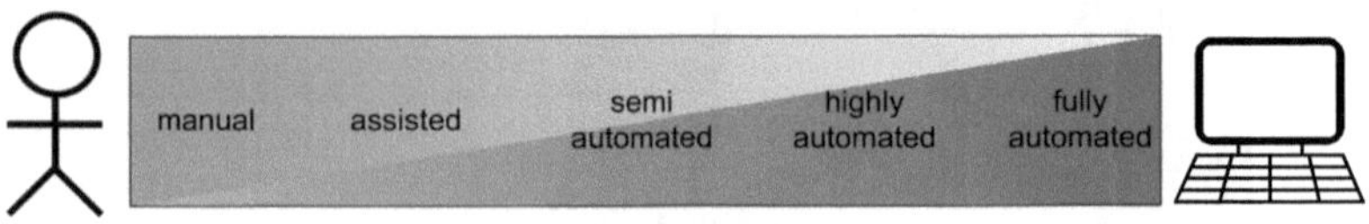

Fig. 5 Assistance and automation scale (Adapted from Flemisch et al. 2008)

high inner and outer compatibility and using cooperative concepts like adaptivity or arbitration, this leads to cooperative (guidance and) control (e.g. Hoc et al. 2006; Biester 2009; Holzmann 2007; Flemisch et al. 2008; Hakuli et al. 2009; Flemisch et al. 2011). One instantiation of cooperative guidance and control is Conduct-by-wire (Winner et al. 2008), where the driver is initiating maneuvers with a special interface on the steering wheel. A natural example of cooperative guidance and control would be two humans going hand in hand, or the relationship between a rider and horse, that was explored as a design metaphor (H-Metaphor, e.g. Flemisch et al. 2003) and implemented and tested as a haptic-multimodal mode of Interaction H-Mode, (e.g. Goodrich et al. 2006; Heesen et al. 2010; Bengler et al. 2011).

In a more abstract perspective, the distribution of control between the human and the automation can be simplified on a 1-dimensional scale of assistance and automation. This scale integrates ADAS systems of 2011 like LKAS and ACC and expands this to higher regions of automation. In highly automated driving, the automation has the capability to drive autonomously at least in limited settings, but the automation is used in a way that the driver is still meaningfully involved in the driving task and in control of the vehicle (e.g. Hoeger et al. 2008). The assistance and automation scale also represents a simple metaphor for the distribution of control between the human and a computer that can be applied to information and display design, as the example in Fig. 5 shows. Overall, the potential, gradual revolution towards automation will bring new challenges for information ergonomics in vehicles.

5 Driver Information Systems: Central Display and Control

In the area of driver information systems, too, a clear increase in functionality and thus of display and control elements can be observed. In order to still guarantee optimum operability for the driver despite this increase, almost all premium manufacturers have introduced concepts of a central display and control (BMW iDrive™, AUDI MMI™, Daimler COMMAND™) in order to facilitate interaction for the driver with the high number of driver information systems (Zeller et al. 2001).

These interaction concepts are characterized by a high degree of menu control. They use a central controller, possibly with haptic feedback information, as well as further input options (handwriting recognition, speech recognition) as a multimodal enhancement (Bernstein et al. 2008; Bengler 2000). Besides visualization, thus aspects of speech and semantic interaction appear as well in the area of human-machine communication (Which commands can be spoken? Which symbol supply is "understandable" for the recipient?) (Bengler 2001; Zobl et al. 2001).

Technologies that are applied in the area of multimodal interaction have been significantly refined with regards to speech, gesture and mimic. On the output side, improved speech synthesis systems, tactile force feedback and complex graphic systems can be applied. This way the user can be given the impression of a dynamic interaction partner with its own personality.

However, driving the car remains the main task. The introduction of new ways of communication (GPRS, UMTS, DAB) accompanies this process: "As a general rule increased functionality normally produces higher complexity for handling" (Haller 2000). This increase in information (parking lots, weather, office, news, entertainment) thus has to be in accordance with driving-related information. The European Communities' recommendation (2006) outlines what determines the quality of an MMI solution for the car. What are primarily required here is minimum distraction from the driving task, an error-resistant interaction design, and intuitive user-controlled operation concepts. Even if future vehicle automation will enable a temporary release from the driving task, the necessity to communicate a higher complexity of task combinations or task transitions will even increase.

5.1 Instrument Cluster and HUD in the Vehicle

In general, the instrument cluster represents the central display for the communication of data required by the driver in order to check the primary driving task and the vehicle's status. In many cases, information is displayed via needle instruments (speed, RPM, oil pressure, tank level) and increasingly via extensive displays with graphic qualities.

Fig. 6 Arrangement of displays and input elements in a BMW as seen from the driver's seat (BMW Group)

Technological development heads into the direction of getting away from mechanical analog instruments towards freely programmable instrument clusters and extensive displays with information about the vehicle and its status.

From a technical point of view, information that is relevant for the driver is processed from measured data for these display instruments, taking the above mentioned maxims into consideration. Partly, this is done by one display/gauge per sensor, partly, however, available in-vehicle data are combined to form integrating information units, in order to generate additional information for the driver or facilitate information processing, e.g. onboard computer data.

The amount of available in-vehicle information is growing continually (e.g. integration of in-vehicle infotainment systems) due to the increase of in vehicle sensor systems and vehicle functionality. At the same time, display elements are increasingly designed as large-size graphic displays, and even a complete combination of instrument cluster and CID (Central Information Display) is conceivable (Fig. 6).

In order to facilitate information reception and keep it as little distracting for the driver as possible also in the future, in-vehicle information thus has to be better structured and processed and summarized to driving relevant and secondary/tertiary information, i.e. information that is non-relevant for the primary driving task. One example for this are current designs of the driver's operation position, especially within the premium segment of the automobile industry.

In addition to this, new display technologies allow for placing information directly within the driver's field of view (head-up display HUD), according to the situation.

The HUD is one approach to present driving-relevant system information with a minimum of visual distraction. Car manufacturers started to integrate HUDs that

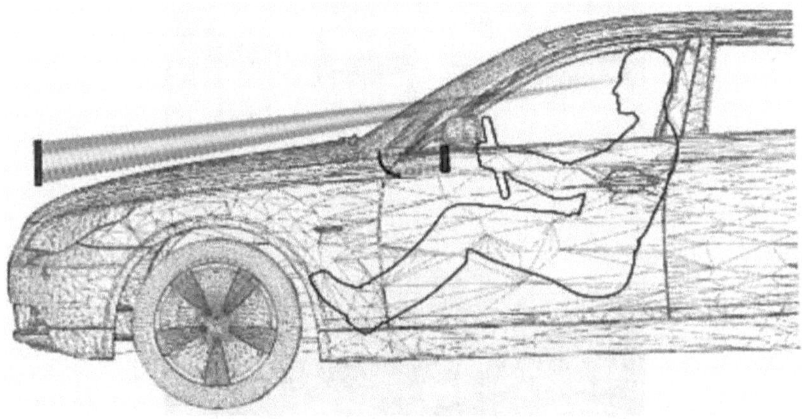

Fig. 7 Design and mode of operation of a conventional HUD in a vehicle (BMW Group)

project information via the windshield to the driver's field of view. The driver can recognize this information with a minimum of visual demand and accommodation.

A HUD is assembled as follows: a display is shown virtually at a larger distance via optical elements (convergence lens, or concave mirror), with the optical element being installed right under the windshield so that its mirror image is perceived as overlaying the road scenario. With HUDs that are common today – being called conventional here – a distance of approximately 2.5–3 m has been chosen. The optic is adjusted in a way that the mirror image of the display appears just above the hood, a field that lies just within the area that does not have to be kept clear from impairments according to the certification specification for vehicles. Through tilting of the first deflection mirror, the display can be adjusted to different levels for the driver's eye. The display distance mentioned is a compromise between the technical optic precision required for the assembly and the cost. As the information is displayed via the bent windshield, the first deflection mirror has to be bent as well in order to avoid double images that would occur in case of differently reflected sight lines from the left and right eye. Double images, which occur in the front and back side of the windshield due to reflection, can be avoided through windows with a wedge-shaped windshield interlayer (see Fig. 7).

As the display space is limited, there should not be too much information on the display. Basically, only information relevant for the driving task should be displayed there. The current speed is displayed constantly, preferably in digital form to avoid confusion of the moving indicator with moving objects outside. It is also useful to display certain status data relevant for driving such as information about whether the ACC or LKAS are switched on, or temporary information like speed limits, no passing, and navigation advice (Fig. 8). It has been proven that, compared to a conventional instrument display, the advantage of a HUD lies in the fact that the driver only needs half the angle (approx. 4° instead of approx.8°) and thus only half the time for averting his/her eyes to receive information. Thus, according to a study by Gengenbach (1997), the overall time of the eye focusing

Fig. 8 Display contents on a conventional HUD (example: BMW HUD) (BMW Group)

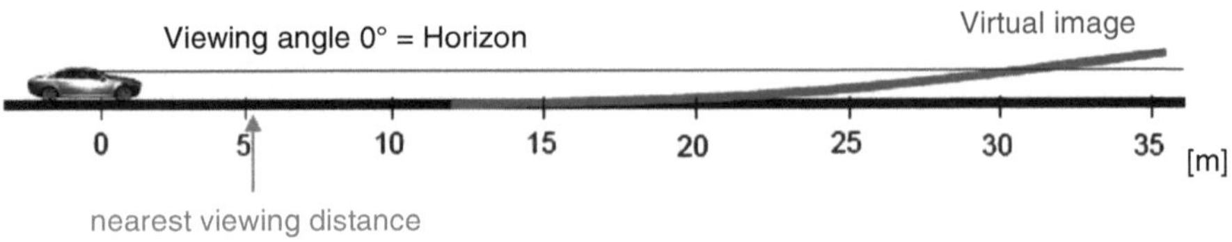

Fig. 9 Virtual plane of a contact analog HUD (cHUD) according to Schneid (2007)

on the speedometer is reduced from approx. 800 ms to approx. 400 ms through the HUD. Furthermore, the route and important changes in other road users' behavior (e.g. lighting up of the preceding vehicle's brake lights) remain within the peripheral field of view.

However, the ergonomic aim is to indicate information that can be directly assigned to a position on the road. A first proposal was already made by Bubb (1981), who indicated the braking distance and alternatively the safety distance by a light bar virtually lying in the calculated distance on the surface of the road. Bergmeier (2009) has made principal investigations and found out that by the use of a virtual plane at a distance of 50 m no assignment error in relation to the road occurs. Schneid (2007) has equipped an AUDI experimental car with a special HUD optic where the virtual plane in the surrounding area really lies on the road. For larger distances, this plane is inclined. In this construction, he has considered the results of Bergmeier (Fig. 9).

Fig. 10 View through the windshield of an experimental car equipped with a contact analog HUD

Figure 10 shows the view of the driver in an AUDI experimental car. In front of the car, a virtual light bar is indicated. The distance of this bar is the safety distance depending on the speed. The bar moves sidewards depending on the steering wheel position and lateral acceleration. It shows the actual future path within 1.2 s. Therefore, it also helps to keep the safe Time to Line Crossing. When the ACC recognizes a leading vehicle, the color of the bar changes to green. Now, the working behavior of the ACC can actually be understood by the user. When the distance becomes too short and the deceleration of the leading car is so high that intervention of the driver is necessary, the color changes to red and the bar would flash in case of emergency. In real driving experiments, Israel et al. (2010) could show that even in case of a more complex indication of this safety distance (e.g. different safety zones through red, orange and green colored surfaces) no negative influence on the driving performance was observed but a significantly reduced workload of the driver. The hedonic and pragmatic quality was also evaluated significantly better than the usual indication of the information in the dashboard.

A further application of the contact analog HUD is the indication of navigation hints. The indication technique allows fixing an arrow on the surface of the street which points in a corner to the road that should be taken. Present ongoing experiments show the great potential of this technique for unambiguous navigation hints.

Simulator experiments have been carried out in order to find out the acceptance of such an ergonomic lay-out of Advanced Driver Assistance Systems

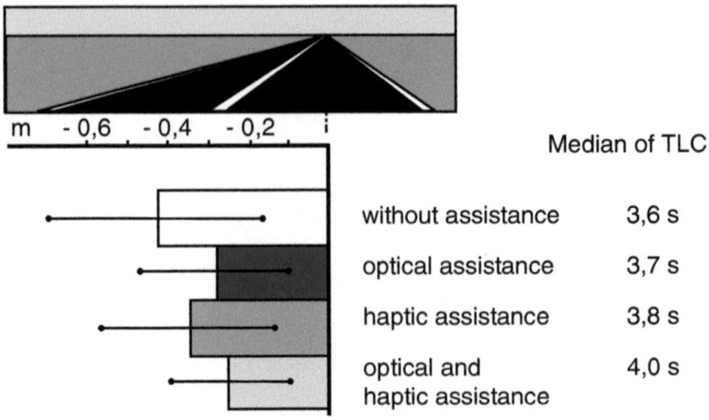

Fig. 11 Deviation from the middle of the path with different ADAS

(Lange 2007)[1]. Figure 11 shows results concerning lane keeping assistance. All results have been compared with the base line behavior. One form of the lay-out was to indicate an impending lane departure only optically in the form of a small arrow next to the safety distance bar. A further lay-out was designed as a merely haptic assistance. When the driver took too little Time to Line Crossing (TLC), this was indicated by a moment on the steering wheel that drew him back to the correct course (Heading Control, HC). The last lay-out combined the haptic and optical indication. The results of this lane departure warning showed that in the base line drive, i.e. the situation without assistance, the driver kept an average position of about 40 cm left of the middle of his/her line. In case of optical assistance, this was reduced to 30 cm. The standard deviation decreased too. In case of haptic assistance, the mean value of deviation to the left increased again. It was lowest when haptic and optical indication were combined. The TTC values are all in a safe range. In case of combination of optical and haptic assistance, those values are significantly higher.

In the same experiment, longitudinal aspects were also observed. In this case, the optical assistance was provided through a speed limit indication in the HUD. Haptic assistance was implemented through a so called active accelerator pedal (AAP). That means, if the driver exceeds the given speed limit, an additional force is applied that informs the driver that he should reduce pressure on the pedal. The results show no difference between the behavior under base line and optical indication conditions (Fig. 12). The average speed excess is reduced by half again in both cases, i.e. haptic and combined haptic and optical indication. The subjective assessment of workload measured by the NASA TLX questionnaire does not show any difference between no indication and optical indication either. It is significantly lower in case of haptic and

[1] Between 25 and 30 experienced drivers participated in the experiments as subjects. In every case, the basic hypotheses were tested by ANOVA and with the Scheffé test. Only results that discriminate better than 5% error probability are reported.

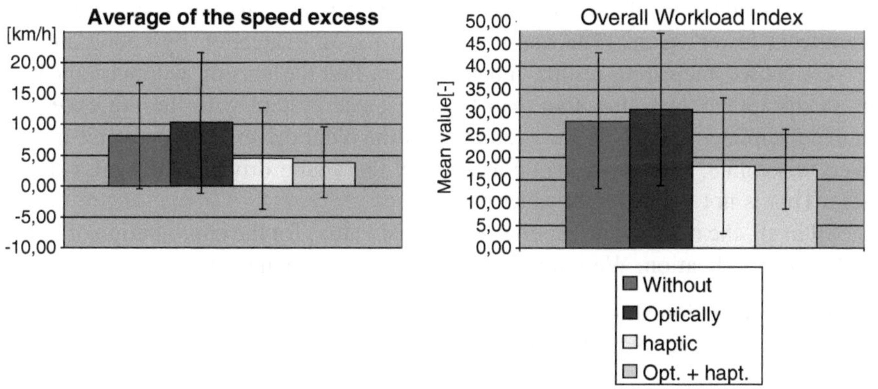

Fig. 12 Speed excess and workload index in connection with different longitudinal ADAS

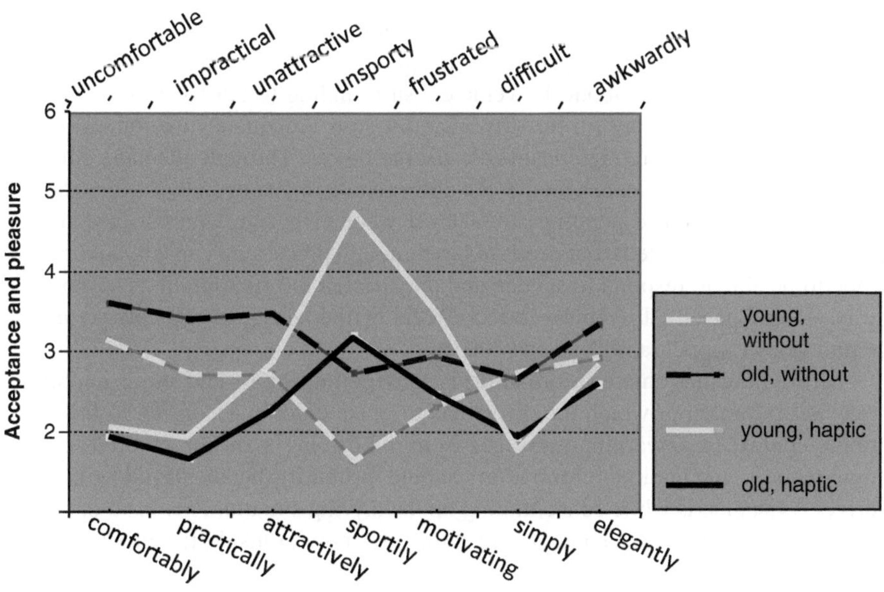

Fig. 13 Subjective evaluation of the different ADAS

combined indication. Looking at the maximum speed excess, no difference between young and old drivers could be observed. Regarding the subjective workload, no difference between young and old people can be observed.

Especially with view to acceptance of assistance systems, the subjective evaluation is of great importance. That was investigated by means of a questionnaire battery of semantic differentials. The results show in general better evaluations for the haptic and the combination of haptic and optical indication (Fig. 13).

Only the evaluation of the question about whether the lay-out is sportily shows significant lower values. The question asking for an evaluation by young and old drivers shows interesting results. Young drivers find the version without indication in all disciplines less attractive than the assistance system with the big exception in the discipline "sportily". The evaluation of the older drivers is even worse here in every discipline. In case of haptic assistance, the young drivers really miss sportiness. This is not valid for the older drivers.

All in all, the results here reported show best values for the combination of haptic and optical indication. We interpreted this in the following way:

- Haptic indication should indicate **what** the driver should do, and
- Optical indication **why** he should do it.

Thus, the question arises as to what information should be presented to the driver when and where/on which display? Besides ergonomic structuring criteria, the presented information has to be divided into information relevant for the driving situation and permanently displayed status information.

What is getting more and more important because of the introduction of advanced driver assistance systems is information that sees the driven vehicle in relation to information about the vehicle's surrounding (other road users, guiding info, etc.) that is relevant for the driver's safety too. Adequate situation awareness will play an even more significant role for the driver. Through adequate situation awareness the driver can become the information manager. He/she can decide according to the corresponding stress level whether he/she wants to accomplish certain tasks (self-paced), but needs to have enough processing capacity at the same time to be able to respond to sudden and unexpected information in an appropriate way (forced-paced). In-vehicle tasks: effects of modality, driving relevance, and redundancy (Seppelt and Wickens 2003).

Partitioning in-vehicle information is thus getting more and more important: relevant information which the driver has to grasp quickly according to the situational context (e.g. warning messages from the driving assistant, control-resumption request, navigation information) should primarily be displayed within the driver's visual axis while concentrating on the driving situation, i.e. head-up; status indications on the other hand should rather be shown outside the driver's visual axis, i.e. head-down.

A more advanced way of presenting information in a more augmenting approach is the so called contact analog HUD (Schneid 2008; Bubb 1986), which displays the HUD's virtual image onto the lane. It allows minimizing the visual demand significantly. On the other hand, it gets more and more relevant to avoid cognitive capture und perceptual tunneling effects that are known from the area of aviation in combination with complex HUD settings. (Milicic et al. 2009; Israel et al. 2010).

5.2 General Design Rules

It is assumed that up to 90% of the information that has to be processed by the driver while driving is restricted to the visual sensory channel (e.g. Rockwell 1972). For a safe drive, the driver's eye should be concentrating on the road, except for short glances into the mirrors and in-vehicle instruments. Thus, particularly in case of a high workload on the driver, using acoustic signals for human–vehicle communication are recommendable (e.g. simple tones, tone sequences, speech; e.g. Wickens 1980, 1984, 2002). Visual information presentation is suitable especially if the information is complex and shown continually and does not require the driver's immediate reaction and/or frequent verification from his/her side (Stevens et al. 2002). In case of visual information presentation (e.g. non-verbal icons or pictographs, menu systems based on speech recognition), too much visual distraction should be avoided through adequate measures.

Concerning the design of human–vehicle communication, first recommendations or guidelines were written in the mid-90es (to get an overview, see Green 2001). The "European Statement of Principles on Human Machine Interface" (Commission of the European Communities 2000) for example formulates recommendations regarding general design, installation, information presentation, interaction with readouts and control elements, system behavior, and the available information about the system (instruction manuals). This document refers for example to the corresponding ISO standards for human–vehicle communication (e.g. EN ISO 15008 2003 for the design of visual information, or EN ISO 15006 2003 for acoustic information design). Those give specific hints on how perceptibility and comprehensibility of the information displayed can be ensured.

Furthermore, there are ISO standards by now for the dialog management of driver information and assistance systems (Transport Information and Control Systems, TICS; EN ISO 15005 2002). These standards introduce the following main criteria for the development and evaluation of an appropriate dialog management:

1. Compatibility with driving, simplicity, and timing or priorities
2. Degree of being self-explanatory, conformity with error expectation and fault tolerance
3. Consistency and controllability

While the first criterion primarily deals with the target of creating an optimum time interleave of dialog management and driving task, the above criteria under (2) and (3) consider particularly content aspects or formulate requirements regarding display and operability of the dialog management. It can be expected that the dialog management thus will be less complex. These design principles can easily be transferred to other areas of human–vehicle communication, even though they were originally formulated for in-vehicle information and assistance systems. Human–vehicle communication is thus to be optimized by taking above mentioned recommendations into consideration.

5.3 Time Aspects of Human–Vehicle Communication

In order to optimize the time structure of human–vehicle communication with the target of an appropriate time interleave of driving and human–vehicle communication, two approaches have to be highlighted in particular and shall be discussed in the following:

1. Moment of presenting various pieces of information.
2. Time structure of one particular piece of information.

5.3.1 Moment of Information Presentation

In contrast to other communication situations, the moment of information presentation and the recipient's time budget when the information is presented and designed in the car play an important role. Following Rasmussen's level model of driving behavior (1980), Lee (2004) names the optimum point in time for information presentation and discusses possible interferences with secondary tasks:

- Strategic level – decisions before driving (e.g. choice of route or time of departure): minutes to days before the situation
- Tactical level – striving for a constantly low risk while being on the road (e.g. anticipation of driving maneuvers, based on knowledge about oneself, the vehicle and expectations): 5–60 s before the situation
- Operational level – direct choice and conduct of maneuvers to avert danger (e.g. change of lanes, or driving around unforeseen obstacles): 0.5–5 s before the situation.

Even though those periods are not stated precisely, the discussion itself makes clear that it is necessary to consider aspects of chronology and duration. The time budget in case of acoustic in-vehicle information presentation is very important as the acoustic channel can only be used serially (i.e. one information at a time). Following Grice (1967/1989, 1975/1993), a cooperative system with acoustic information presentation is thus characterized by presentation of the appropriate message at the right point in time (maxim of relation; see Sect. 2.2). Assuming for example that the spoken presentation of a navigation hint takes 4–5 s, and interpretation and motoric translation into action take another 6–8 s, the information thus should be presented 13 s before the desired action. This can lead to problems if there are many tasks in a row that need to be fulfilled in a short period of time.

Schraagen (1993) thus recommends integrating not more than two commands in one message. Streeter et al. (1985) state, too, that there should be a break of appropriate length between the presentation of two different pieces of information to give the driver a chance, for example, to repeat the message he/she just heard. In a similar way, Stanton and Baber (1997) advise against using speech signals as warning messages because of their length. In this case, redundant solutions are to be preferred (e.g. acoustic signals combined with written texts).

5.3.2 Time Structure of Information Presentation

In addition to that, dependency on presentation modalities needs to be considered in order to structure messages time-wise. Following this other aspects are important in case of a combined speech and visual information presentation than in case of speech and acoustic information presentation.

For example, speech-based visual information in menu-controlled information systems results in different stress effects for the driver, depending on the amount of information presented simultaneously on the in-vehicle display. Studies that were carried out in other contexts but the vehicle context, mostly recommend using menu structures with 7 +/– 2 options per menu level (e.g. Norman 1991), or avoiding structures that involve more than three menu levels (e.g. Miller 1981). The decisive factor, however, is certainly not the mere number of options alone but also their cohesion with regards to content, or how they can be grouped. It was shown, for example, that an increasing number of levels can lead to more problems for the system users orienting themselves (i.e. the users do not know where they presently are in the menu; Paap and Cooke 1997), or that system functions with ambiguous terminology lead to a loss in control performance, especially with several menu levels (Miller and Remington 2002). Those results, however, cannot be simply transferred to the vehicle context, where the driver has to operate such menu-controlled information systems while driving. Rauch et al. (2004) showed that – regarding both objective performance in accomplishing the driving task and menu operation, and workload perceived by the driver – menu-controlled information systems are more recommendable if they do have several menu levels, but if each level contains only few menu options. It seems as if those systems challenge the driver less visually: due to fact that the driver needs to look at the display only for short and homogeneous time periods, these variants of information presentation can be connected with the driving task more easily.

In a study by Totzke et al. (2003b), menu variants that display only parts of the information system prove to be better (e.g. only the menu level that is currently used by the driver). Consequently, Norman's recommendation (1991) of a menu system design which keeps the user informed about his/her position within the menu system does not seem to apply to the vehicle context. One possibility would be, for example, to display not only the current level or menu option, but also further information about hierarchically higher or lower menu levels and options. What seems to be decisive is the driver's eye glance behavior. In agreement with this opinion, other authors state that glances that take too long (e.g. more than 2 s; Zwahlen et al. 1988) or too many glances into the interior of the vehicle (up to 40 glances per operation task with unexperienced drivers; Rockwell 1988) are to be seen critically (for an overview, see Bruckmayr and Reker 1994). Consequently, general interruptibility of the information reception and interaction is one of the most important requirements for the design. That means, the driver should be able to interrupt the interaction at any time and resume it at a later point in time without greater effort or "damage" (Keinath et al. 2001).

Similar recommendations are made for time aspects of acoustic in-vehicle information. Simpson and Marchionda-Frost (1984) showed e.g. that the speed of information presentation in a pilot's warning system has significant influence on reaction time and comprehensibility of the information. Likewise, Llaneras, Lerner, Dingus and Moyer (2000) report less decrease in the driving performance and longer reaction times for secondary tasks in case of faster information presentation in the vehicle context.

After all, in order to optimize the time structure of acoustic in-vehicle information, a variation in the ratio of speech pauses seems to be suitable following results regarding "natural" human speech behavior: longer pauses lead to easier structuring of the presented information for the listener which means that the additional load on the driver due to communication could be reduced. However, up to now in-vehicle information and communication systems, like acoustic information presentation in navigation and route guidance for example, make use of this possibility only to a limited extent. Hence, pauses within the information presented by the vehicle are currently due to the linkage algorithms of the corresponding vehicle system and not due to semantics. Regarding a possible speech input by the driver, which establishes itself in the vehicle context and enables the driver to use vehicle functions via speech recognition, one can also expect that the driver will soon take on the time pattern of the system. For example, slower feedback from the system results in slower speech tempo of the user. Literature refers to this phenomenon as the so called Lee effect (Langenmayr 1997) or as "convergence" (Giles and Powisland 1975). Furthermore, it was observed that users of a speech-controlled system often react to recognition errors of the system with a louder and more accentuated way of speaking (so called Lombard effect; Langenmayr 1997). This indicates that the user associates recognition errors with background noise or assumes an insufficient signal-to-noise ratio, whereas the real reason is often that the system either responds too early or too late (Bengler et al. 2002). These findings clearly indicate that system users (here: drivers) use natural behavior of human-human communication in human-machine communication (here: human–vehicle communication) too. The target thus should be to avoid the occurrence of the Lombard effect due to high recognition rates as far as possible and minimize its impacts technically and, if necessary, through adequate dialog control.

6 Summary: Final Remarks

This chapter described the marginal conditions for information and communication in the situational context "vehicle". Respective approaches in the area of "information" and "communication" were taken into consideration and used to illustrate the fact that the topic information and communication gets more complex within this particular context. Thus, for communication one has to consider general physical-physiological-technical effect aspects ("How are signals that are needed for transporting information from the transmitter to the recipient conveyed?"), psychological effect aspects ("Under which circumstances will the recipient perceive and process the signals?"), and semantic effect aspects ("Which coherence needs to be there between information and signal so that the recipient can comprehend the adequate information sent out by the signal?"). In the vehicle context, however, time aspects and the necessary interruptibility of communication play a further role that is of utmost importance. Accordingly, time-related effect aspects of communication also exist in the vehicle ("When should information be presented by the transmitter in order to have the recipient react adequately at the right point in time?").

The theoretical background of the requirement for interruptibility of information and communication in the vehicle is based on the assumption that human processing capacity is limited, i.e. it is not possible to perceive and process an unlimited amount of information at the same time and also react adequately. If those available information processing capacities are exceeded, the acting person's performance decreases. Accordingly, the necessity arises to optimize information and communication in the vehicle context, so that only a minimum of driving and safety relevant negative effects on the driving performance arise.

Currently, assumptions of the multiple resources model, enhanced by a so called bottleneck on the processing level of reaction selection (i.e. translation of the encoded stimuli into precise action orders), are to be supported. This results in a dependency on alertness regarding the selection of different alternatives for reaction. Particularly strong interferences between driving and information and communication can be expected for attention-related control processes on the knowledge level. One has to consider, however, that during driving the driver is exposed to a high number of different information sources and communication partners which have to be processed via different processing channels or interact with the driver in varying speech ratios. A taxonomy, which was introduced in this context, attempts to line out the complexity of in-vehicle information and communication.

At first, it was suggested to optimize the time structure of human–vehicle communication with the target of creating an adequate interleave of driving and communication.

1. To this end, the information output would have to be sequenced and introduced early enough before a relevant situation.
2. Eventually, the creation of a time structure within an information situation was discussed. This should be done, for example, by adapting the information presented to the time requirements within the vehicle context via amount (visual information) or adequate pauses (acoustic information).

Furthermore, different alternatives were discussed regarding the optimization of a content-wise structure of human–vehicle communication with the target of reducing its complexity.

1. Initially, the phrase structure of such systems was highlighted, using the example of standard navigation systems and menu-controlled driver assistance systems.
2. Then, the significance of semantic aspects (content structure and topic) and syntactic aspects of communication situations was presented.
3. Finally, the information content or redundancy of information was addressed, which can be varied, e.g., via removal of speech elements, via additional use of icons and pictographs, or tones and tone sequences, or via combination of sensory channels for the information output.

In-vehicle communication and information thus means more than the use of a cell phone or the internet while driving and represent a heterogeneous field of requirements that the driver is exposed to. The basis of an optimum design of human–vehicle communication should be features of in-vehicle human-human communication, i.e. features of small talk with the passenger: such conversations illustrate how communication and information and driving can run off simultaneously without causing problems.

The prerequisite is an adequate realization of physical features of visual or acoustic information that is presented by the vehicle. For this purpose, corresponding guidelines or ISO/DIN standards were named as a reference.

References

Aguiló, M. C. F. (2004). Development of guidelines for in-vehicle information presentation: Text vs. speech. Dissertation, Blacksburg, VA: Virginia Polytechnic Institute and State University.

Alm, H., & Nilsson, L. (1994). Changes in driver behaviour as a function of handsfree mobile phones-a simulator study. *Accident, Analysis and Prevention, 26*, 441–451.

Alm, H., & Nilsson, L. (1995). The effects of a mobile telephone task on driver behaviour in a car following situation. *Accident Analysis and Prevention, 27*, 707–715.

Arnett, J. J., Offer, D., & Fine, M. A. (1997). Reckless driving in the adolescence: "State" and "trait" factors. *Accident Analysis and Prevention, 29*, 57–63.

Atchley, P., & Dressel, J. (2004). Conversation limits the functional field of view. *Human Factors, 46*(4), 664–673.

Aylett, M., & Turk, A. (2004). The smooth signal redundancy hypothesis: A functional explanation for relationships between redundancy, prosodic prominence, and duration in spontaneous speech. *Language and Speech, 47*(1), 31–56.

Bader, M. (1996). *Sprachverstehen. Syntax und Prosodie beim Lesen*. Opladen: Westdeutscher Verlag.

Bahrick, H. P., & Shelly, C. (1958). Time sharing as an index of automatization. *Journal of Experimental Psychology, 56*(3), 288–293.

Baxter, J. S., Manstead, A. S. R., Stradling, S. G., Campbell, K. A., Reason, J. T., & Parker, D. (1990). Social facilitation and driver behaviour. *British Journal of Psychology, 81*, 351–360.

Becker, S., Brockmann, M., Bruckmayr, E., Hofmann, O., Krause, R., Mertens, A., Nin, R., & Sonntag, J. (1995). *Telefonieren am Steuer* (Berichte der Bundesanstalt für Straßenwesen. Mensch und Sicherheit, Vol. M 45). Bremerhaven: Wirtschaftsverlag NW.

Belz, S. M., Robinson, G. S., & Casali, J. G. (1999). A new class of auditory warning signals for complex systems: Auditory icons. *Human Factors, 41*, 608–618.

Bengler, K. (2000). Automotive speech-recognition – Success conditions beyond recognition rate. *Proceedings LREC 2000*, Athens, pp. 1357–1360.

Bengler, K. (2001). Aspekte der multimodalen Bedienung und Anzeige im Automobil. In Th Jürgensohn & K.-P. Timpe (Eds.), *Kraftfahrzeugführung* (pp. 195–206). Berlin/Heidelberg/ New York: Springer.

Bengler, K., Geutner, P., Niedermaier, B., & Steffens, F. (2000). "Eyes free–Hands free" "oder "Zeit der Stille". Ein Demonstrator zur multimodalen Bedienung im Automobil, DGLR-Bericht 2000–02 "Multimodale Interaktion im Bereich der Fahrzeug und Prozessführung" (S.299–307). Bonn: Deutsche Gesellschaft für Luft- und Raumfahrt e.V. (DGLR).

Bengler, K., Noszko, Th., & Neuss, R. (2002). Usability of multimodal human-machine-interaction while driving. *Proceedings of the 9-th World Congress on ITS [CD-Rom], 9th World Congress on Intelligent Transport Systems*, ITS America, 10/2002.

Bernstein, A., Bader, B., Bengler, K., & Kuenzner, H. (2008). Visual-haptic interfaces in car design at BMW. In Martin Grunwald (Ed.), *Human haptic perception, basics and applications*. Berlin/Basel/Boston: Birkhäuser.

Seppelt, B., &. Wickens C. D. (2003). Technical report AHFD-03-16/GM-03-2 August 2003.

Bolte, U. (1991). Das aktive Stellteil – ein ergonomisches Bedienkonzept. Fortschrittsberichte VDI-Reihe 17 Nr. 75.

Boomer, D. S. (1978). The phonemic clause: Speech unit in human communication. In A. W. Siegman & S. Feldstein (Eds.), *Multichannel integrations of nonverbal behavior* (pp. 245–262). Hillsdale: Erlbaum.

Boomer, D. S., & Dittmann, A. T. (1962). Hesitation pauses and juncture pauses in speech. *Language and Speech, 5*, 215–220.

Braune, R., & Wickens, C. D. (1986). Time-sharing revisited: Test of a componential model for the assessment of individual differences. *Ergonomics, 29*(11), 1399–1414.

Brems, D. J., Rabin, M. D., & Waggett, J. L. (1995). Using natural language conventions in the user interface design of automatic speech recognition system. *Human Factors, 37*(2), 265–282.

Briem, V., & Hedman, L. R. (1995). Behavioral effects of mobile telephone use during simulated driving. *Ergonomics, 38*, 2536–2562.

Broadbent, D. E. (1958). *Perception and communication*. London: Pergamon.

Brookhuis, K. A., de Vries, G., & de Waard, D. (1991). The effects of mobile telephoning on driving performance. *Accident, Analysis and Prevention, 23*, 309–316.

Bruckmayr, E., & Reker, K. (1994). Neue Informationstechniken im Kraftfahrzeug – Eine Quelle der Ablenkung und der informatorischen Überlastung? *Zeitschrift für Verkehrssicherheit, 40*(1), 12–23.

Bühler, K. (1934). *Sprachtheorie – Die Darstellungsfunktion der Sprache*. Stuttgart: Fischer.

Byblow, W. D. (1990). Effects of redundancy in the comparison of speech and pictorial display in the cockpit environment. *Applied Ergonomics, 21*(2), 121–128.

Byblow, W. D., & Corlett, J. T. (1989). Effects of linguistic redundancy and coded-voice warnings on system response time. *Applied Ergonomics, 20*(2), 105–108.

Carlson, K., & Cooper, R. E. (1974). A preliminary investigation of risk behaviour in the real world. *Personality and Social Psychology Bulletin, 1*, 7–9.

Charlton, S. G. (2004). Perceptional and attentional effects on drivers' speed selection at curves. *Accident Analysis and Prevention, 36*, 877–884.

Commission of the European Communities. (2000). Commission recommendation of 21 December 1999 on safe and efficient in-vehicle information and communication systems: A European statement of principles on human machine interface. *Official Journal of the European Communities*, 25.01.2000.

Consiglio, W., Driscoll, P., Witte, M., & Berg, W. P. (2003). Effects of cellular telephone conversations and other potential interference on reaction time in a braking response. *Accident Analysis and Prevention, 35*, 495–500.

Cooper, P. J., Zheng, Y., Richard, C., Vavrik, J., Heinrichs, B., & Siegmund, G. (2003). The impact of hands-free message reception/response on driving task performance. *Accident Analysis and Prevention, 35*, 23–53.

DePaulo, B., & Rosenthal, R. (1988). Ambivalence, discrepancy and deception in nonverbal communication. In R. Rosenthal (Ed.), *Skill in nonverbal communication* (pp. 204–248). Cambridge: Oelschlager, Gunn & Hain.

Deutsch, J. A., & Deutsch, D. (1963). Attention: Some theoretical considerations. *Psychological Review, 70*, 80–90.

Dingus, T. A., & Hulse, M. C. (1993). Some human factors design issues and recommendations for automobile navigation information systems. *Transportation Research, 1C*(2), 119–131.

Ebbesen, E. B., & Haney, M. (1973). Flirting with death: Variables affecting risk-taking at intersections. *Journal of Applied Psychology, 3*, 303–324.

Eckstein, L. (2001). Entwicklung und Überprüfung eines Bedienkonzepts und von Algorithmen zum Fahren eines Kraftfahrzeugs mit aktiven Sidesticks. Fortschrittsberichte VDI-Reihe 12, Nr. 471.

Einhorn, H. J., Kleinmuntz, D. N., & Kleinmuntz, B. (1979). Linear regression and process-tracing models of judgement. *Psychological Review, 86*, 465–485.

Ellinghaus, D., & Schlag, B. (2001). *Beifahrer – Eine Untersuchung über die psychologischen und soziologischen Aspekte des Zusammenspiels von Fahrer und Beifahrer.* Köln: UNIROYAL-Verkehrsuntersuchung.

EN ISO 15005. (2002). *Straßenfahrzeuge – Ergonomische Aspekte von Fahrerinformations- und -assistenzsystemen – Grundsätze und Prüfverfahren des Dialogmanagements.* Berlin: Beuth.

EN ISO 15006. (2003). Ergonomische Aspekte von Fahrerinformations- und Assistenzsystemen – Anforderungen und Konformitätsverfahren für die Ausgabe auditiver Informationen im Fahrzeug. Berlin: DIN Deutsches Institut für Normung e.V.

EN ISO 15008. (2003). Ergonomische Aspekte von Fahrerinformations- und Assistenzsystemen – Anforderungen und Bewertungsmethoden der visuellen Informationsdarstellung im Fahrzeug. Berlin: DIN Deutsches Institut für Normung e.V.

Endsley, M. R. (1995). Toward a theory of situation awareness in dynamic systems. *Human Factors, 37*(1), 32–64.

Evans, L. (1991). *Traffic safety and the driver.* New York: Van Nostrand Reinhold.

Färber, B., & Färber, B. (1999). *Telematik-Systeme und Verkehrssicherheit* (Berichte der Bundesanstalt für Straßenwesen, Reihe Mensch und Sicherheit, Heft M 104). Bergisch Gladbach: Wirtschaftsverlag NW.

Fassnacht, G. (1971). Empirische Untersuchung zur Lernkurve. Unveröffentlichte Dissertation. Bern: Philosophisch-Historische Fakultät der Universität Bern.

Flemisch, F. O., Adams, C. A., Conway, S. R., Goodrich, K. H., Palmer, M. T., & Schutte, P. C. (2003). *The H-metaphor as a guideline for vehicle automation and interaction* (NASA/TM-2003-212672). Hampton: NASA, Langley Research Center.

Flemisch, F. O., Kelsch, J., Löper, C., Schieben, A., Schindler, J., & Heesen, M. (2008). *Cooperative control and active interfaces for vehicle assistance and automation.* München: FISITA.

Gardiner, M. M. (1987). *Applying cognitive psychology to user-interface design.* Chichester: Wiley.

Garland, D. J., Wise, J. A., & Hopkin, V. D. (1999). *Handbook of aviation human factors.* Mahwah: Erlbaum.

Giles, H., & Powisland, P. F. (1975). *Speech styles and social evaluation.* London: Academic.

Gilliland, S. W., & Schmitt, N. (1993). Information redundancy and decision behavior: A process tracing investigation. *Organizational Behavior and Human Decision Processes, 54*, 157–180.

Goldman-Eisler, F. (1968). *Psycholinguistics: Experiments in spontaneous speech.* New York: Academic.

Goldman-Eisler, F. (1972). Pauses, clauses, sentences. *Language and Speech, 15*, 103–113.

Green, P. (2001, February). Variations in task performance between younger and older drivers: UMTRI research on telematics. *Paper presented at the association for the advancement of automotive medicine conference on aging and driving*, Southfield.

Green, P., & Williams, M. (1992). Perspective in orientation/navigation displays: A human factors test. *The 3rd international conference on vehicle navigation systems and information systems*, Oslo, pp. 221–226.

Green, P., Hoekstra, E., & Williams, M. (1993). Further on-the-road tests of driver inter-faces: Examination of a route guidance system and a car phone (UMTRI-93-35). Ann Arbor, MI: University of Michigan Transportation Research Institute.

Grice, H. P. (1967/1989). Logic and conversation. In H. P. Grice (Ed.), *Studies in the way of words*. Cambridge: Harvard University Press.

Grice, H. P. (1975/1993). Logik und Konversation. In G. Meggle (Ed.), *Handlung, Kommunikation, Bedeutung – Mit einem Anhang zur Taschenbuchausgabe* (pp. 243–265). Frankfurt: Suhrkamp.

Grosjean, F., & Deschamps, A. (1975). Analyse contrastive des variables temporelles des l'anglais et du francais. *Phonetica, 31*, 144–184.

Haigney, D., & Westerman, S. J. (2001). Mobile (cellular) phone use and driving: A critical review of research methodology. *Ergonomics, 44*(82), 132–143.

Haigney, D., Taylor, R. G., & Westerman, S. J. (2000). Concurrent mobile (cellular) phone use and driving performance: Task demand characteristics and compensatory processes. *Transportation Research Part F, 3*, 113–121.

Hakuli, S., et al. (2009). Kooperative automation. In H. Winner, S. Hakuli, & G. Wolf (Eds.), *Handbuch Fahrerassistenzsysteme*. Wiesbaden: Vieweg+Teubner Verlag.

Hall, J. F. (1982). *An invitation to learning and memory*. Boston: Allyn & Bacon.

Hanks, W., Driggs, X. A., Lindsay, G. B., & Merrill, R. M. (1999). An examination of common coping strategies used to combat driver fatigue. *Journal of American College Health, 48*(3), 135–137.

Hanowski, R., Kantowitz, B., & Tijerina, L. (1995). *Final report workload assessment of in-cab text message system and cellular phone use by heavy vehicle drivers in a part-task driving simulator*. Columbus: Battelle Memorial Institute.

Helleberg, J. R., & Wickens, C. D. (2003). Effects of data-link modality and display redundancy on pilot performance: An attentional perspective. *International Journal of Aviation Psychology, 13*(3), 189–210.

Hoeger, R., Amditis A., Kunert M., Hoess, A., Flemisch, F., Krueger, H.-P., Bartels, A., & Beutner, A. (2008). Highly automated vehicles for intelligent transport: HAVEit approach; *ITS World Congress*, 2008.

Irwin, M., Fitzgerald, C., & Berg, W. P. (2000). Effect of the intensity of wireless telephone conversations on reaction time in a braking response. *Perceptual and Motor Skills, 90*, 1130–1134.

Israel, B., Seitz, M., & Bubb, H. (2010). Contact analog displays for the ACC – A conflict of objects between stimulation and distraction. *Proceedings of the 3-rd international conference on applied human factors and ergonomics [CD-Rom]*, Miami.

Kahneman, D. (1973). *Attention and effort*. Englewood Cliffs: Prentice Hall.

Kames, A. J. (1978). A study of the effects of mobile telephone use and control unit design on driving performance. *IEEE Transactions on Vehicular Technology, VT-27*(4), 282–287.

Kantowitz, B. H. (1995). Simulator evaluation of heavy-vehicle driver workload. *Paper presented at the human factors and ergonomics society 39th annual meeting*, Santa Monica.

Keele, S. W. (1973). *Attention and human performance*. Pacific Palisades: Goodyear.

Keinath, A., Baumann, M., Gelau, C., Bengler, K., & Krems, J. F. (2001). Occlusion as a technique for evaluating in-car displays. In D. Harris (Ed.), *Engineering psychology and cognitive ergonomics* (Vol. 5, pp. 391–397). Brookfield: Ashgate.

Kienle, M., Damböck, D., Kelsch, J., Flemisch, F., & Bengler, K. (2009). Towards an H-mode for highly automated vehicles: Driving with side sticks. *Paper presented at the Automotive UI 2009*, Essen.

Kinney, G. C., Marsetta, M., & Showman, D. J. (1966). *Studies in display symbol legibility, Part XXI. The legibility of alphanumeric symbols for digitized television (ESD-TR-66-177)*. Bedford: The Mitre Corporation.

Kircher, A., Vogel, K., Törnros, J., Bolling, A., Nilsson, L., Patten, C., Malmström, T., & Ceci, R. (2004). *Mobile Telefone simulator study (VTI Meddelande 969A)*. Sweden: Swedish National Road and Transport Research Institute.

Krallmann, D., & Ziemann, A. (2001). *Grundkurs Kommunikationswissenschaft*. München: Wilhelm Fink Verlag.

Krüger, H.-P. (1989). Speech chronemics – a hidden dimension of speech. Theoretical background, measurement, and clinical validity. *Pharmacopsychiatry, 22*(Supplement), 5–12.

Krüger, H.-P., & Vollrath, M. (1996). Temporal analysis of speech patterns in the real world using the LOGOPORT. In J. Fahrenberg & M. Myrtek (Eds.), *Ambulatory assessment* (pp. 101–113). Seattle: Hogrefe & Huber.

Krüger, H.-P., Braun, P., Kazenwadel, J., Reiß, J. A., & Vollrath, M. (1998). *Soziales Umfeld, Alkohol und junge Fahrer* (Berichte der Bundesanstalt für Straßenwesen, Mensch und Sicherheit, Vol. M 88). Bremerhaven: Wirtschaftsverlag.

Kryter, K. D. (1994). *The handbook of hearing and the effects of noise*. San Diego: Academic.

Labiale, G. (1990). In-car road information: Comparisons of auditory and visual presentations. *Proceedings of the human factors and ergonomics society 34th annual meeting*, Human and Factors Ergonomics Society, Santa Monica, pp. 623–627.

Lager, F., & Berg, E. (1995). *Telefonieren beim Autofahren*. Wien: Kuratorium für Verkehrssicherheit, Institut für Verkehrspsychologie.

Lamble, D., Kauranen, T., Laakso, M., & Summala, H. (1999). Cognitive load and detection thresholds in car following situations: Safety implications for using mobile (cellular) telephones while driving. *Accident Analysis and Prevention, 31*, 617–623.

Langenmayr, A. (1997). *Sprachpsychologie*. Göttingen: Hogrefe.

Lasswell, H. D. (1964). The structure and function of communication in society. In L. Bryson (Ed.), *The communication of Ideas* (pp. 37–51). New York: Institute for Religious and Social Studies.

Lee, J. D. (2004). Preface of the special section on driver distraction. *Human Factors, 46*(4), 583–586.

Liu, Y.-C., Schreiner, C. S., & Dingus, T. A. (1999). Development of human factors guidelines for advanced traveller information systems (ATIS) and commercial vehicle operations (CVO): Human factors evaluation of the effectiveness of multi-modality displays in advanced traveller information systems. FHWA-RD-96-150. Georgetown: U.S. Department of Transportation, Federal Highway Administration.

Llaneras, R., Lerner, N., Dingus, T., & Moyer, J. (2000). Attention demand of IVIS auditory displays: An on-road study under freeway environments. *Proceedings of the international ergonomics association 14th congress*, Human Factors and Ergonomics Society, Santa Monica, pp. 3-238–3-241.

Loftus, G., Dark, V., & Williams, D. (1979). Short-term memory factors in ground controller/pilot communication. *Human Factors, 21*, 169–181.

Maag, C., & Krüger, H.-P. (im Druck). Kapitel in der Enzyklopädie

Macdonald, W. A., & Hoffmann, E. R. (1991). Drivers' awareness of traffic sign information. *Ergonomics, 41*(5), 585–612.

Maletzke, G. (1998). *Kommunikationswissenschaft im Überblick – Grundlagen, Probleme, Perspektiven*. Opladen/Wiesbaden: Westdeutscher Verlag.

Matthews, R., Legg, S., & Charlton, S. (2003). The effect of phone type on drivers subjective workload during concurrent driving and conversing. *Accident Analysis and Prevention, 35*, 451–457.

McKnight, A. J., & McKnight, A. S. (1993). The effect of cellular phone use upon driver attention. *Accident Analysis & Prevention, 25*, 259–265.

Merten, K. (1993). Die Entbehrlichkeit des Kommunikationsbegriffs. In G. Bentele & M. Rühl (Eds.), *Theorien öffentlicher Kommunikation* (pp. 188–201). München: Ölschläger.

Metker, T., & Fallbrock, M. (1994). Die Funktionen der Beifahrerin oder des Beifahrers für ältere Autofahrer. In U. Tränkle (Ed.), *Autofahren im Alter* (pp. 215–229). Köln: Deutscher Psychologen Verlag.

Meyer, D. E., & Kieras, D. E. (1997). A computational theory of executive cognitive processes and multiple-task performance: Part 1, basic mechanisms. *Psychological Review, 104*(1), 3–65.

Milicic, N., Platten, F., Schwalm, M., & Bengler, K. (2009). Head-up display und das situationsbewusstsein. *VDIBERICHT 2009, 2085*, 205–220.
Miller, D. P. (1981). The depth/breadth tradeoff in hierarchical computer menus. *Proceedings of the human factors society – 25th annual meeting*, Human Factors Society, Santa Monica, pp. 296–300.
Miller, C. G., & Remington, R. W. (2002). A computational model of web navigation: Exploring interactions between hierarchical depth and link ambiguity [Web-Dokument]. http://www.tri.sbc.com/hfweb/miller/hfweb.css. [03.10.2002].
Moray, N. (1967). Where is capacity limited? A survey and a model. *Acta Psychologica, 27*, 84–92.
Morrow, D., Lee, A., & Rodvold, M. (1993). Analysis of problems in routine controller pilot-communication. *International Journal of Aviation Psychology, 3*(4), 285–302.
Mutschler, H., Baum, W., & Waschulewski, H. (2000). Report of evaluation results in German (Forschungsbericht, SENECA D22-3.1).
Neale, V. L., Dingus, T. A., Schroeder, A. D., Zellers, S., & Reinach, S. (1999). Development of human factors guidelines for advanced traveler information systems (ATIS) and commercial vehicle operations (CVO): Investigation of user stereotypes and preferences (Report No. FHWA-RD-96-149). Georgetown: U.S. Department of Transportation, Federal Highway Administration.
Nilsson, L. (1993). Behavioural research in an advanced driving simulator: Experiences of the VTI system. *Paper presented at the proceedings of the human factors and ergonomics society 37th annual meeting*, Human Factors and Ergonomics Society, Santa Monica, pp. 612–616.
Nilsson, L., & Alm, H. (1991). Elderly people and mobile telephone use – effects on driver behaviour? *Proceedings of the conference strategic highway research program and traffic safety on two continents*, Gothenburg, pp. 117–133.
Norman, D. A. (1981). Categorization of action slips. *Psychological Review, 88*(1), 1–15.
Norman, D. A. (1983). Some observations on mental models. In D. Gentner & A. L. Stevens (Eds.), *Mental models* (pp. 7–14). Hillsdale: Erlbaum.
Norman, D. A. (1986). New views of information processing: Implications for intelligent decision support systems. In E. Hollnagel, G. Mancini, & D. D. Woods (Eds.), *Intelligent decision support in process environment* (pp. 123–136). Berlin: Springer.
Norman, K. L. (1991). *The psychology of menu selection: Designing cognitive control of the human/computer interface*. Norwood: Ablex Publishing Corporation.
Noszko, Th., & Zimmer, A. (2002). Dialog design für sprachliche und multimodale Mensch-Maschine-Interaktionen im Automobil. BDP Kongress 2002, Regensburg.
Noyes, J. M., & Frankish, C. R. (1994). Errors and error correction in automatic speech recognition systems. *Ergonomics, 37*(11), 1943–1957.
Nunes, L., & Recarte, M. A. (2002). Cognitive demands of hands-free-phone conversation while driving. *Transportation Research Part F, 5*, 133–144.
Oviatt, S. (2000). Multimodal system processing in mobile environments, to appear in *Proceedings of 6th international conference on spoken language processing* (ICSLP 2000), Beijing, 16–20 Oct 2000.
Paap, K., & Cooke, N. (1997). Design of menus. In M. Helander, T. Landauer, & P. Prabhu (Eds.), *Handbook of human–computer interaction* (pp. 533–572). Amsterdam: Elsevier.
Parkes, A. M. (1991). Drivers business decision making ability whilst using carphones. In E. J. Lovesey (Ed.), *Contemporary ergonomics, proceedings of the ergonomic society annual conference*. London: Taylor & Francis.
Parkes, A., & Coleman, G. (1993). Route guidance systems: A comparison of methods of presenting directional information to the driver. In E. J. Lovesey (Ed.), *Contemporary ergonomics. Proceedings of the ergonomic society 1990. Annual conference*. London: Taylor and Francis.

Parkes, A. M., & Hooijmeijer, V. (2001). The influence of the use of mobile phones on driver situation awareness (Web-Dokument). http://www-nrd.nhtsa.dot.gov/departments/nrd-13/driver-distraction/PDF/2.PDF. [04.05.04].

Pashler, H. (1994). Dual-task interference in simple tasks: Data and theory. *Psychological Bulletin, 116*, 220–244.

Patten, C. J. D., Kircher, A., Östlund, J., & Nilsson, L. (2004). Using mobile telephones: Cognitive workload and attention resource allocation. *Accident Analysis and Prevention, 36*, 341–450.

Penka, A. (2000). Vergleichende Untersuchung zu Fahrerassistenzsystemen mit unterschiedlichen aktiven Bedienelementen. Technische Universität München. Garching 2000.

Power, M. J. (1983). Are there cognitive rhythms in speech? *Language and Speech, 26*(3), 253–261.

Praxenthaler, M., & Bengler, K. (2002). Zur Unterbrechbarkeit von Bedienvorgängen während zeitkritischer Fahrsituationen. BDP-Kongress 2002, Regensburg.

Rasmussen, J. (1980). *Some trends in man–machine-interface design for industrial process plants (Technical Report Riso-M-2228)*. Riso: Riso National Laboratory.

Rasmussen, J. (1983). Skills, rules, knowledge: Signals signs and symbols and other distinctions in human performance models. *IEEE Transactions: Systems, Man & Cybernetics, SMC-13*, 257–267.

Rauch, N., Totzke, I., & Krüger, H.-P. (2004). Kompetenzerwerb für Fahrerinformationssysteme: Bedeutung von Bedienkontext und Menüstruktur. In VDI Gesellschaft Fahrzeug- und Verkehrstechnik (Ed.), *Integrierte Sicherheit und Fahrerassistenzsysteme* (VDI-Berichte, Vol. 1864, pp. 303–322). Düsseldorf: VDI-Verlag.

Reason, J. T. (1979). Actions not as planned: The price of automation. In G. Underwood & R. Stevens (Eds.), *Aspects of consciousness* (pp. 67–89). London: Academic.

Reason, J. T. (1986). Recurrent errors in process environments: Some implications for intelligent decision support systems. In E. Hollnagel, G. Mancini, & D. Woods (Eds.), *Intelligent decision support in process environment* (NATO ASI Series 21st ed., pp. 255–271). Berlin: Springer.

Redelmeier, D. A., & Tibshirani, R. J. (1997). Association between cellular-telephone calls and motor vehicle collisions. *The New England Journal of Medicine, 336*, 453–458.

Reed, M. P., & Green, P. A. (1999). Comparison of driving performance on-road and in a low-cost simulator using a concurrent telephone. *Ergonomics, 42*(8), 1015–1037.

Reiss, J. A. (1997). *Das Unfallrisiko mit Beifahrern – Modifikatoren in Labor und Feld*. Aachen: Shaker Verlag.

Reynolds, A., & Paivio, A. (1968). Cognitive and emotional determinants of speech. *Canadian Journal of Psychology, 22*(3), 161–175.

Rockwell, T. H. (1972). Eye movement analysis of visual information acquisition in driving – An overview. *Proceedings of the 6th conference of the Australian road research board,* Australian Road Research Board, Canberra, pp. 316–331.

Rockwell, T. H. (1988). Spare visual capacity in driving revisited: New empirical results for an old idea. In A. G. Gale, C. M. Haslegrave, S. P. Taylor, P. Smith, & M. H. Freeman (Eds.), *Vision in vehicles II* (pp. 317–324). Amsterdam: North Holland Press.

Rohmert, W., & Breuer, B. (1993). Driver behaviour and strain with special regard to the itinerary; Prometheus Report Phase III; Technische Hochschule Darmstadt, 1993.

Rolls, G., & Ingham, R. (1992). *"Safe" and "unsafe" – A comparative study of younger male drivers*. Basingstoke/Hampshire: AA Foundation for Road Safety Research, Fanum House.

Rolls, G., Hall, R. D., Ingham, R., & McDonald, M. (1991). *Accident risk and behavioural patterns of younger drivers*. Basingstoke/Hampshire: AA Foundation for Road Safety Research, Fanum House.

Rueda-Domingo, T., Lardelli-Claret, P., de Dios Luna-del-Castillo, J., Jiménez-Moleón, J. J., Garcia-Martin, M., & Bueno-Cavanillas, A. (2004). The influence of passengers on the risk of the driver causing a car collision in Spain analysis of collisions from 1990 to 1999. *Accident Analysis and Prevention, 36*, 481–489.

Schlag, B., & Schupp, A. (1998). Die Anwesenheit Anderer und das Unfallrisiko. *Zeitschrift für Verkehrssicherheit, 44*(2), 67–73.
Schmidtke, H. (1993). *Lehrbuch der Ergonomie* (3rd ed.). München: Hanser Verlag.
Schmitt, N., & Dudycha, A. L. (1975). Positive and negative cue redundancy in multiple cue probability learning. *Memory and Cognition, 3*, 78–84.
Schneid, M. (2008). Entwicklung und Erprobung eines kontaktanalogen Head-up-Displays im Fahrzeug, Dissertation Lehrstuhl für Ergonomie TUM. Munchen, Technical University.
Schneider, V. I., Healy, A. F., & Barshi, I. (2004). Effects of instruction modality and readback on accuracy in following navigation commands. *Journal of Experimental Psychology: Applied, 10* (4), 245–257.
Schnotz, W. (1994). *Aufbau von Wissensstrukturen. Untersuchungen zur Kohärenzbildung beim Wissenserwerb mit Texten.* Weinheim: Beltz PsychologieVerlagsUnion.
Schraagen, J. M. C. (1993). Information processing in in-car navigation systems. In A. M. Parkes & S. Franzen (Eds.), *Driving future vehicles* (pp. 171–186). Washington, DC: Taylor & Francis.
Searle, J. R. (1982). *Ausdruck und Bedeutung – Untersuchungen zur Sprechakttheorie*. Frankfurt/ Main: Suhrkamp.
Shannon, C. E., & Weaver, W. (1949). *The mathematical theory of communication*. Urbana: The University of Illinois Press.
Shinar, D., Tractinsky, N., & Compton, R. (2005). Effects of practice, age, and task demands, on interference from a phone task while driving. *Accident, Analysis and Prevention, 37*, 315–326.
Shneiderman, B. (1998). *Designing the user interface*. Reading: Addison, Wesley & Longman.
Simpson, C. A. (1976). Effects of linguistic redundancy on pilots comprehension of synthesized speech. *Proceedings of the Twelfth Annual Conference on Manual Control* (NASA CR-3300). NASA – Ames Research Center, Moffett Field.
Simpson, C. A., & Marchionda-Frost, K. (1984). Synthesized speech rate and pitch effects on intelligibility of warning messages for pilots. *Human Factors, 26*(5), 509–517.
Sperandio, J. C., & Dessaigne, M. F. (1988). Une comparisation expérimentale entre modes de présentation visuels ou auditifs de messages d'informations routières a des conducteurs automobiles. *Travail Humain, 51*, 257–269.
Stanton, N. A., & Baber, C. (1997). Comparing speech vs. text displays for alarm handling. *Ergonomics, 40*(11), 1240–1254.
Stein, A. C., Parseghian, Z., & Allen, R. W. (1987). A simulator study of the safety implications of cellular mobile phone use (Paper No. 405 (March)). Hawthorne, CA: Systems Technology, Inc.
Steinbuch, K. (1977). *Kommunikationstechnik*. Berlin: Springer.
Stevens, A., Quimby, A., Board, A., Kersloot, T., & Burns, P. (2002). Design guidelines for safety of in-vehicle information systems (Project report PA3721/01). Workingham: Transport Local Government
Strayer, D. L., & Johnston, W. A. (2001). Driven to distraction: Dual-task studies of simulated driving and conversing on a cellular phone. *Psychological Science, 12*(6), 462–466.
Streeter, L., Vitello, D., & Wonsiewicz, S. A. (1985). How to tell people where to go: Comparing navigational aids. *International Journal of Man Machine Studies, 22*, 549–562.
Tijerina, L., Kiger, S., Rockwell, T. H., & Tornow, C. E. (1995). Final report – Workload assessment of in-cab text message system and cellular phone use by heavy vehicle drivers on the road (Contract No. DTNH22-91-07003). (DOT HS 808 467 (7A)). Washington, DC: U.S. Department of Transportation, NHTSA.
Totzke, I., Meilinger, T., & Krüger, H.-P. (2003a). Erlernbarkeit von Menüsystemen im Fahrzeug – mehr als "nur" eine Lernkurve. In Verein Deutscher Ingenieure (Ed.), *Der Fahrer im 21. Jahrhundert* (VDI-Berichte, Vol. 1768, pp. 171–195). Düsseldorf: VDI-Verlag.
Totzke, I., Schmidt, G., & Krüger, H.-P. (2003b). Mentale Modelle von Menüsystemen – Bedeutung kognitiver Repräsentationen für den Kompetenzerwerb. In M. Grandt (Ed.), *Entscheidungsunterstützung für die Fahrzeug- und Prozessführung* (DGLR-Bericht 2003–04, pp. 133–158). Bonn: Deutsche Gesellschaft für Luft- und Raumfahrt e.V.

Treffner, P. J., & Barrett, R. (2004). Hands-free mobile phone speech while driving degrades coordination and control. *Transportation Research Part F, 7*, 229–246.

Ungerer, D. (2001). Telefonieren beim Autofahren. In K. Hecht, O. König, & H.-P. Scherf (Eds.), *Emotioneller Stress durch Überforderung und Unterforderung*. Berlin: Schibri-Verlag.

Verwey, W. B. (1993). How can we prevent overload of the driver? In A. M. Parkes & S. Franzén (Eds.), *Driving future vehicles* (pp. 235–244). London: Taylor & Francis.

Verwey, W. B., & Veltman, J. A. (1995). Measuring workload peaks while driving. A comparison of nine common workload assessment techniques. *Journal of Experimental Psychology: Applied, 2*, 270–285.

Violanti, J. M., & Marshall, J. R. (1996). Cellular phones and traffic accidents: an epidemiological approach. *Accident Analysis and Prevention, 28*, 265–270.

Vollrath, M., & Totzke, I. (2000). In-vehicle communication and driving: An attempt to overcome their interference [Web-Dokument]. Driver distraction internet forum. http://www-nrd.nhtsa.dot.gov/departments/nrd-13/driver-distraction/Papers.htm. [22.10.2001].

Vollrath, M., Meilinger, T., & Krüger, H.-P. (2002). How the presence of passengers influences the risk of a collision with another vehicle. *Accident Analysis and Prevention, 34*(5), 649–654.

Watzlawick, P., Beavin, J. H., & Jackson, D. D. (1996). *Menschliche Kommunikation* (9 unv Auflageth ed.). Bern: Verlag Hans Huber.

Welford, A. T. (1952). The psychological refractory period and the timing of high-speed performance – A review and a theory. *British Journal of Psychology, 43*, 2–19.

Welford, A. T. (1967). Single channel operation in the brain. *Acta Psychologica, 27*, 5–21.

Wickens, C. D. (1980). The structure of attentional resources. In R. Nickerson (Ed.), *Attention and performance* (Vol. VIII, pp. 239–257). Englewood Cliffs: Erlbaum.

Wickens, C. D. (1984). Processing resources in attention. In R. Parasuraman (Ed.), *Varieties of attention* (pp. 63–102). London: Academic.

Wickens, C. D. (2002). Multiple resources and performance prediction. *Theoretical Issues in Ergonomics Science, 3*(2), 159–177.

Wickens, C. D., & Hollands, J. (2000). *Engineering psychology and human performance* (3rd ed.). Upper Saddle River: Prentice Hall.

Winner, H., Hakuli, S. et al. (2006). Conduct-by-Wire – ein neues Paradigma für die Weiterentwicklung der Fahrerassistenz. Workshop Fahrerassistenzsysteme 2006, Löwenstein.

Zeller, A., Wagner, A., & Spreng, M. (2001). iDrive – Zentrale Bedienung im neuen 7er von BMW. In *Elektronik im Kraftfahrzeug* (VDI-Berichte, Vol. 1646, pp. 971–1009). Düsseldorf: VDI-Verlag.

Zobl, M., Geiger, M., Bengler, K., & Lang, M. (2001). A usability study on hand gesture controlled operation of in-car devices. HCI 2001 New Orleans.

Zwahlen, H. T., Adams, J. C. C., & Schwartz, P. J. (1988). Safety aspects of cellular telephones in automobiles (Paper No. 88058). In *Paper presented at the proceedings of the ISA-TA Conference*, Florence.

Civil Aviation

Jochen Kaiser and Christoph Vernaleken

1 Information Exchange in Civil Aviation

Compared to other transport systems (e.g. automobile and railway transport), the network integration and the exchange of information within civil aviation are highly developed. The information exchange does not only occur between pilots and air traffic controllers as key participants, but also includes other airline stakeholders, e.g. operations control and maintenance as well as airports and passengers. In addition to these information interfaces within the air transport system, there are external interfaces for information exchange, e.g. ground or maritime transport systems.

This results in an extensive information system, which will be briefly described prior to a detailed discussion of selected aspects of information systems in civil aviation from an information ergonomics perspective. It should be noted that this chapter takes an aircraft-centric view on the subject, whereas a comprehensive discussion of Air Traffic Control (ATC) aspects can be found in another chapter.

An aspect of information exchange in civil aviation that is getting more and more important is safety. The relevant information that is needed for a safe flight can be classified into flight safety (information to avoid collisions or extreme weather) and aviation security (information exchange between airlines and national authorities to enhance border security).

J. Kaiser (✉)
Overall Systems Design & Architecture, Cassidian, Manching, Germany
e-mail: InfoErgonomics@me.com

C. Vernaleken
Human Factors Engineering, Cassidian, Manching, Germany

M. Stein and P. Sandl (eds.), *Information Ergonomics*,
DOI 10.1007/978-3-642-25841-1_5,

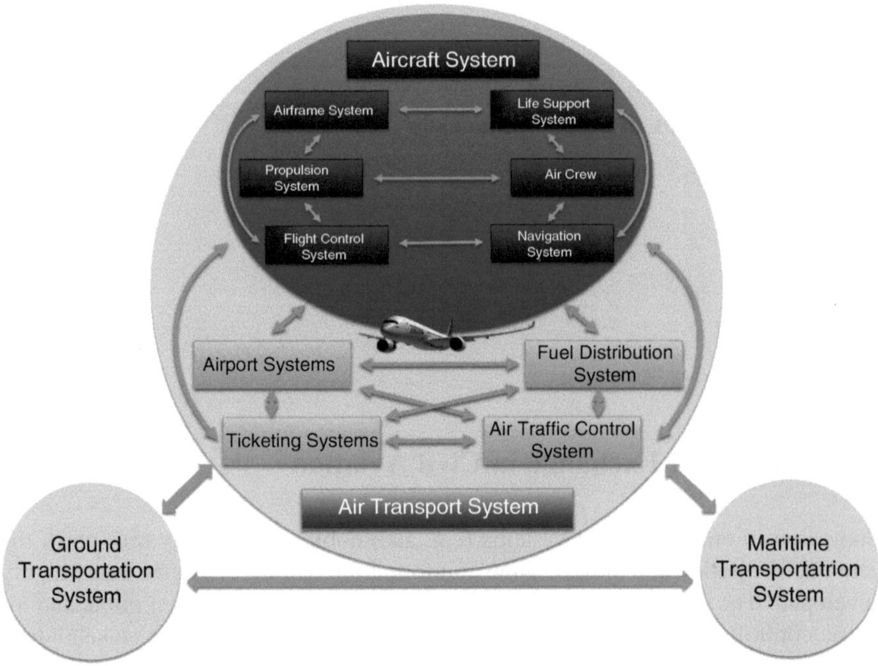

Fig. 1 Schematic representation of the aircraft and transport system-of-systems and the associated information exchange (Based on INCOSE Systems Engineering Handbook (2010))

1.1 *Information Network in Civil Aviation*

The first part of this chapter gives a general overview of the different facets of information flow in civil aviation. The emphasis will be placed on commercial air transport by airlines, not on general aviation. The information network in general aviation adheres to the same principles, but is significantly less complex, and can thus be considered as a subset of the air transport aviation network.

A first step in developing a picture of the civil aviation information network is the identification of the involved participants and their relations. Participants may be persons (e.g. flight crews), system elements (e.g. aircraft) and external systems (e.g. other transport systems in the ground or maritime environment). Figure 1 only provides a simplified overview of the different participants and their relationship regarding information exchange.

Airlines contribute the majority of the system elements with their aircraft. The airline users of the information network are first of all the cockpit and cabin crews, in the background, however, every airline has a closely interconnected ground personnel network. Maintenance crews use network information for the planning of aircraft technical services and for communicating with the airworthiness authorities. The ground operation of airlines provides flight planning and handling of passengers, cargo and baggage.

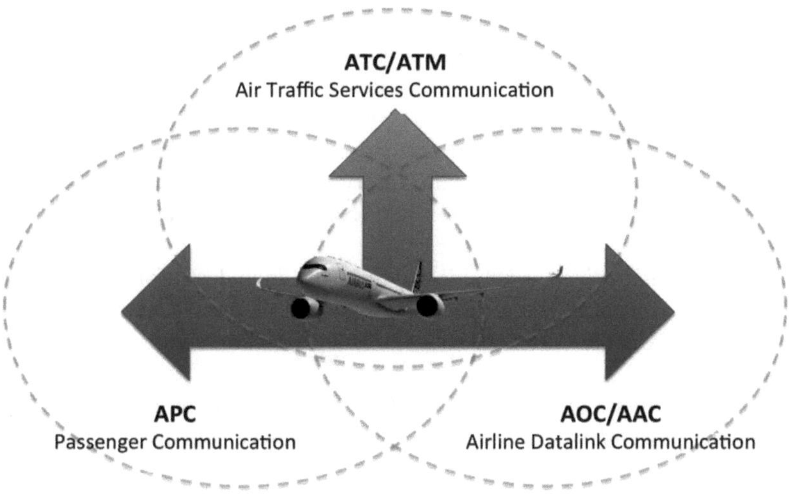

Fig. 2 Major aeronautical data link communications (according to Airbus (2004a))

The passengers themselves represent a part of the information network, as they may use the available data from the airlines and airports to plan their travels in conjunction with other means of transport. Similar to the tracking of baggage, constant information about the position of cargo is used in many fields of industry and economy.

Another fundamental participant in the aviation network is ATC, which will be dealt with in a separate chapter of this book.

The airports with their teams for passenger/cargo support and handling of air vehicles (e.g. fuel, push-back) are seamlessly integrated into the aviation network. Information is also needed for the supply of consumables and meals to the aircraft. Last but not least security (police, customs) and safety (fire and emergency service) are highly dependent on information from the aviation network.

The exchange of information within the air transport system typically takes place via voice, but there has been a clear trend towards data link communication in recent years. In virtually all cases, the driving factor for such transitions was the need to decrease frequency congestion and communication workload. The latter also encompasses the automation of routine status messages, such as automated position in areas inaccessible to radar surveillance without flight crew interaction, or the downlink of maintenance data.

The three major aeronautical data link communication categories in Fig. 2 can be distinguished by the three end systems (Airbus 2004a):

- Air Traffic Services Communications (between aircraft and ATC centres)
- Airline Data Link Communications (AOC & AAC)
- Aeronautical Passenger Communications (APC)

Air Traffic Control/Management (ATC/ATM) encompasses the digital exchange of instructions and clearances between flight deck and controller on the ground, using the Controller Pilot Data Link Communications (CPDLC), Automatic Dependent Surveillance (ADS), i.e. automated position reporting, and Data Link Flight Information Services (FIS).

Airline Operations Control (AOC) communications involves the information exchange between the aircraft and the airline operational centre or operational staff at the airport associated with the safety and regularity of flights. Examples for this information are Notices to Airmen (NOTAM), weather information, flight status, etc.

Airline Administrative Communications (AAC) includes information regarding administrative aspects of the airline business such as crew scheduling and cabin provisioning. Examples are passenger lists, catering requirements, baggage handling, etc.

Airline Passenger Correspondence (APC) includes communication services that are offered to passengers and comprises e-mail, Internet, television, telephony, etc.

1.2 History of Information Exchange in Civil Aviation

In the early days of aviation, communication between pilots and personnel on the ground almost exclusively relied on visual signals, and, as a consequence, the use of flags, light signals and flares was ubiquitous. In civil aviation, telecommunication over great distances was first established by telegraph systems connecting airfields. Since the advent of radio communication, the Morse code has been commonly used in aeronautics to this day, e.g. for the identification of ground-based navigation aids (VOR, DME, NDB, etc.). Radios allowing voice communication were introduced in the late 1920s and are still the main means of exchanging information in civil aviation (Airbus 2004a).

The use of data link was pioneered in the domain of AOC communications by the Airborne Communications Addressing and Reporting System (ACARS), which was introduced in 1978 using a low-speed Very High Frequency (VHF) data link and a character-based transmission protocol limited to a bandwidth of 2.4 kBit/s, which is also referred to as VHF Digital Link (VDL) Mode A (Airbus 2004). ACARS was originally developed in response to a requirement by Piedmont airlines to find a better way of measuring flight crew duty times. As the system matured, organisations found that other time-sensitive information could be transmitted and received through the ACARS system. Today, ACARS is a standard means for airlines to distribute company messages between their aircraft and their respective home bases. Maintenance services also found an important asset in the downlink of real-time aircraft parameters. Today, over 6,000 aircraft are using ACARS worldwide on a daily basis, with more than 20 million messages per month.

1.3 Information Exchange on Board the Aircraft

The structure of the information exchange in the operation of airlines is quite similar and has a close interdependency with the surrounding actors such as

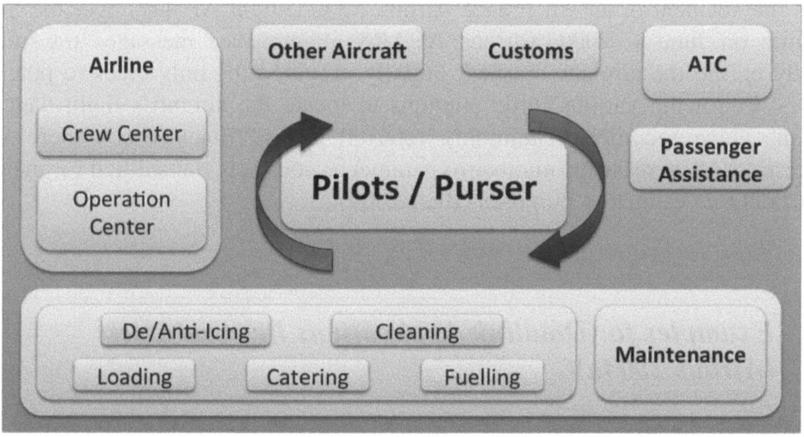

Fig. 3 Information exchange between pilots, purser and other actors (according to Airbus (2004a))

ATC, ground handling, fuel service, etc. A good coordination between all these entities is a fundamental factor for achieving a save and efficient air transport.

Today, the main actors involved in the aircraft operation, aside of the crew, are airline flight operations, airline crew management, maintenance, ground handling, passenger assistance, ATC and services for catering, fuelling and cleaning as indicated in Fig. 3. The exchange of information can be covered either by voice or data link.

The basic information exchange via data link usually results in the implementation of four main applications in airlines that represent more than 50% of the data link traffic (Airbus 2004a):

OOOI – The OOOI (Out, Off, On, In) messages are automatic movement reports used to track aircraft movements, flight progress and delays. The OOOI messages are automatically sent to the airline operation centre, triggered by sensors on the aircraft (such as doors, breaks, gears, etc.).

Weather – METAR and TAF (airport weather information) can be received via ACARS and printed on the cockpit printer. This application allows the pilot to get updated weather conditions and forecasts at the departure and destination airports as well as at other airports along the route. SIGMETs (SIGnificant METeorological advisories) regarding significant meteorological conditions that could affect the flight are distributed for the on-route weather information.

Free Text – On board the aircraft the pilot has the possibility of sending so-called "Free Text" messages to the ground. These messages can be used for miscellaneous communications and can be addressed to maintenance, engineering, dispatch, flight operation, crew management, Air Traffic Management, or even to other traffic.

Maintenance – Aircraft systems status and engine data can be downlinked in real-time from a central on-board maintenance system in order to optimise unscheduled maintenance events. This results in an extension of service intervals without reducing the airworthiness.

In the accident of an Air France Airbus A330 (Flight AF 447) over the South Atlantic on June 1, 2009, various ACARS maintenance messages transmitted shortly before the airplane crashed, initially provided the only clues to potential causes. Following various futile attempts to locate the aircraft's flight data and cockpit voice recorders, commonly referred to as "Black Box", there were suggestions to increase the number of parameters routinely transmitted for incident or accident investigation purposes (BEA 2009).

1.4 Examples for Datalink Applications Per End Users (Airbus 2004a)

1.4.1 Cabin Crew

Information concerning the cabin or the passengers along with the ground handling coordination can be transmitted via radio or data link. Figure 4 illustrates this cabin crew message exchange.

For instance, the cabin technical status, engineering defects or any emergency equipment irregularities or malfunctions can be transmitted by data link to the airline maintenance in real time, so that they can be quickly eliminated and do not affect the aircraft dispatch.

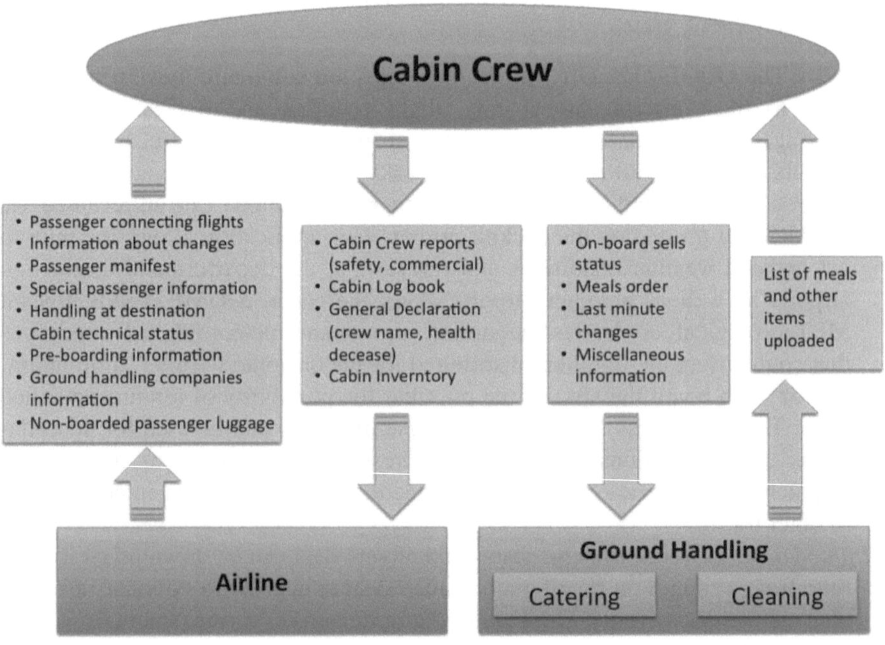

Fig. 4 Cabin crew message exchange (according to Airbus (2004a))

Data link also ensures efficient communication with ground staff (catering or cleaning). In case of modifications of the on-board passenger numbers or specificities, new flight meals (or other catering loads) can be ordered and loaded on board the aircraft.

The following information can be sent via the Cabin Management Terminal (CMT) in the galley area:

- Updated crew list
- Cabin information list (passenger list, route and destination info)
- Credit card authorisation
- Wheel chair request
- On-board sales status

Concerning the administrative tasks of the purser, the cabin logbook can be transmitted directly to the airline at the end of flight legs.

Information concerning passengers can be uplinked from the departure station, for instance if their baggage was not loaded on board in the case of a short connection. Messages to individual passengers are printed and delivered by hand. If the departure was significantly delayed, the stations system also sends a preliminary update about passengers that have missed their original connections and therefore have already been re-booked at this stage. It is possible to report passenger medical problems and obtain advice from the ground.

1.4.2 Passengers

With the increasing available data rates, data link will have an increasing impact on passengers during a flight. It is assumed that passengers will be able to use upcoming wideband satellite data link for e-mail, access to corporate networks via internet, transfer of large files (audio and video images), e-commerce transactions along with Internet, mobile phone and WLAN. The implementation of such services has already started but remains limited due to the limited communication network capacity and data rate. DSP and telecommunication service providers such as Inmarsat are expected to increase their service offers in the near future.

1.5 Excursion: Reuse of Aeronautical Information

Due to the public provision of non-safety-critical aeronautical data via the Internet, today several possibilities of usage of this information can be identified. This was not the original intention at the time the information was generated. One example is the flight tracking of commercial aircraft available in the Internet. Information about departure, arrival and flight status is currently available to airports, passengers and third parties.

A more private or hobby-oriented use of aeronautical data is the integration of air traffic information from ADS-B or transponder into PC-based flight simulations (e.g. Microsoft Flight Simulator).

The information is also reused for interfacing with other transport systems, e.g. coordination with railway schedules and for online tracking of the position of baggage or freight.

1.6 Future Development

The amount of information that will be exchanged will increase in the near future due to new technologies and upcoming air traffic management procedures such as SESAR in Europe and NextGen in the USA. Despite the information ergonomics the handling of this large amount of data will already be a technological challenge, e.g. due to limited frequencies.

One of the main objectives of the Single European Sky ATM Research Programme (SESAR) is the introduction of a System Wide Information Management (SWIM) in Europe. SWIM is a new information infrastructure, which will connect all Air Traffic Management (ATM) stakeholders, aircraft as well as ground facilities. The aim of this new ATM Intranet is to allow seamless information interchange for improved decision-making and the capability of finding the most appropriate source of information while complying with the information security requirements. Through SWIM, all air traffic partners will contemporaneously have the same appropriate and relevant operational picture for their own business needs. The main benefits which are expected are the improvement of aviation safety, the reduction of passenger delays and unnecessary in-flight rerouting or ground taxiing. SESAR is planned to be deployed in several steps and shall be operational around 2020.

2 Technologies for Information Exchange with the Flight Deck

2.1 Exchange and Indication of Aeronautical Information in Commercial Aircraft

2.1.1 Flight Preparation and Planning

Dissemination of Aeronautical Information

Airline operations take place in an environment that is subject to rapid change, with conditions varying on a daily or hourly basis, or even more frequently. Obviously, weather has a major share in these operational variations, but the air traffic infrastructure itself is not static, either (Vernaleken 2011). To cope with these changes, all relevant aeronautical information, irrespective of whether it is distributed electronically or on paper, is updated every 28 days at fixed common dates agreed within the framework of Aeronautical Information Regulation and Control (AIRAC). As a consequence, significant operational changes can only become effective at these previously fixed dates, and the corresponding information must be available at the airline (or any other recipient) at least 28 days in advance of the effective date. Furthermore, it is recommended that all major changes are distributed 56 days in advance.

However, it is not possible to cover all operationally relevant changes through this system. In the airport environment, for example, construction work for maintaining installations, refurbishing traffic-worn pavements or improving and expanding the aerodrome often takes place in parallel to normal flight operations. To minimize impact on traffic throughput, both scheduled maintenance and urgent repair activities are often carried out during night hours or whenever the demand for flight operations is low. It is unrealistic to assume that all of this work can always be planned weeks ahead, since it might become necessary on very short notice (equipment failure, etc.).

As a consequence, all short-term and temporary changes[1] not persisting for the full 28 day period are explicitly excluded from the AIRAC distribution and must be handled differently. The same applies to operationally relevant permanent changes occurring between AIRAC effective dates (ICAO 2004).

Current Handling of Short-Term and Temporary Changes

Today, information on short-term changes and temporary changes to infrastructure or operational procedures is typically conveyed by publication via the following services, depending on the timeframe of validity and the available ahead-warning time (shortest to longest):

- *Air Traffic Control (ATC)*: Short-term, tactical notification of the crew, either via conventional Radio Telephony (R/T) or Controller Pilot Data Link Communications (CPDLC). This service is commonly used to relate information on changes immediately after they have occurred, i.e. when pilots have no other way of knowing.
- *Automatic Terminal Information Service (ATIS)*: Continuous service providing information on the operational status of an aerodrome, such as the runways currently in use for take-off and landing, and prevailing meteorological conditions. ATIS service, which is updated whenever significant changes occur, can either be provided as a pre-recorded voice transmission on a dedicated frequency, via data link (D-ATIS), or both.
- *Notice to Airmen (NOTAM)*[2]: Notification of all relevant short-term changes (temporary or permanent), typically as Pre-flight Information Bulletin (PIB)

---

[1] Short-term changes are alterations that occur between AIRAC effective dates; the changes may be either of temporary or permanent nature.

[2] According to ICAO Annex 15, NOTAM should preferably be distributed by means of telecommunications, but a list of all valid NOTAM in plain text must be provided in print on a monthly basis. NOTAM transmitted via telecommunications are usually referred to as Class I, whereas those distributed in print are Class II.

during the pre-flight briefing (ICAO 2004). Unless they relate to equipment failures and other unforeseen events, NOTAM are typically released several hours before the changes they announce become effective. Information conveyed by NOTAM may be valid for several hours, days or weeks, and is thus mainly used at a strategic level during flight planning.
- *Aeronautical Information Publication (AIP) Supplements*: temporary changes of longer duration (3 months and beyond), e.g. runway closure for several months due to pavement replacement.

If temporary changes are valid for 3 months or longer, they are published as an AIP Supplement. Alternatively, in case extensive text and/or graphics is required to describe information with shorter validity, this will also require an AIP Supplement. AIP supplements often replace NOTAM and then reference the corresponding serial number of the NOTAM (ICAO 2004).

The information from the services above, the first three of which are routinely directly available to flight crews, is complementary and partially redundant, since e. g. runway closure information might be reported by all of the services listed above. NOTAM information, however, forms the strategic baseline for flight planning, flight preparation and decision-making, and therefore deserves special attention.

With the exception of some AIP Supplements, all of these services provide textual or verbal information only.

2.2 On-board Traffic Surveillance and Alerting System

2.2.1 Airborne Collision Avoidance System (ACAS)

To prevent airborne mid-air collisions, all turbine-engine aircraft with a maximum certified take-off mass in excess of 5,700 kg or authorisation to carry more than 19 passengers must be equipped with an Airborne Collision Avoidance System (ACAS) (ICAO 2001b).

ACAS is a system based on air traffic control transponder technology. It interrogates the transponders of aircraft in the vicinity, analyses the replies to determine potential collision threats, and generates a presentation of the surrounding traffic for the flight crew. If other aircraft are determined to be a threat, appropriate alerts are provided to the flight crew to assure separation. Currently, there are three ACAS versions: ACAS I is limited to the visualisation of proximate traffic and alerting to potential collision threats, whereas ACAS II is additionally capable of providing vertical escape manoeuvres to avert a collision. ACAS III is a projected future system offering horizontal escape manoeuvres as well.

Currently, the only ACAS implementation fulfilling the criteria of the ICAO ACAS mandate is the Traffic Alert and Collision Avoidance System (TCAS II) as described in RTCA DO-185A (RTCA 1997). It requires a Mode S transponder, but offers some backward compatibility with the older Mode A/C transponders.

Therefore, the discussion in subsequent sections focuses on TCAS II as the current state-of-the-art.

2.2.2 Traffic Surveillance and Conflict Detection

Each Mode S transponder has a unique ICAO 24 bit address, which serves two purposes, i.e. an unambiguous identification and the possibility of addressing communication to a specific aircraft. Interrogations from secondary surveillance radars (SSR) are transmitted on the 1,030 MHz uplink frequency, whereas transponder replies and other transmissions are made on the 1,090 MHz downlink frequency.

TCAS uses the Mode S and Mode C interrogation replies of other aircraft to calculate bearing, range and closure rate for each so-called 'intruder'. By combining this data with altitude information from transponder replies, where available, three-dimensional tracks for at least 30 surrounding aircraft are established. Vertical speed estimates are derived from the change in altitude (RTCA 1997). These tracks are then used to create a representation of the surrounding traffic for the flight crew and analysed to determine potential collision threats.

Depending on aircraft configuration and external conditions, TCAS detection is typically limited to a range of 30–40 NM and encompasses a maximum vertical range of ±9,900 ft (±3,000 m) above and below ownship (Airbus 2005). TCAS ranges are fairly accurate and usually determined with errors less than 50 m. However, the bearing error may be quite large: for elevation angles within ±10°, the maximum permissible bearing error is 27°, and it may be as high as 45° for higher elevation values (RTCA 1997).

By extrapolating the acquired tracks for the surrounding traffic, TCAS calculates the Closest Point of Approach (CPA) and the estimated time τ before reaching this point for each intruder. Alerts are generated if the CPA for an intruder is within a certain time (τ) and altitude threshold such that it might represent a collision hazard. For aircraft not reporting altitude, it is assumed that they are flying at the same altitude as ownship.

TCAS features two alert levels. It typically first generates a so-called Traffic Advisory (TA), a caution alert that there is potentially conflicting traffic. If the collision threat becomes real, TCAS determines an appropriate vertical evasive manoeuvre and triggers a Resolution Advisory (RA) warning, which identifies the intruder causing the conflict as well as the vertical manoeuvre and vertical rate required to resolve the conflict. In certain situations, depending on the geometry of the encounter or the quality and age of the vertical track data, an RA may be delayed or not selected at all. RAs are only generated for intruders reporting altitude.

The collision avoidance logic used in determining nature and strength of the vertical evasive manoeuvre for Resolution Advisories is very complex due to the numerous special conditions. Additional complexity originates from the fact that

TCAS is capable of handling multiple threat encounters. An exhaustive description of the TCAS logic can be found in Volume 2 of DO-185A (RTCA 1997).

For intruders identified as threats, an appropriate Resolution Advisory evasive manoeuvre is determined, depending on the precise geometry of the encounter and a variety of other factors, such as range/altitude thresholds and aircraft performance. The evasive manoeuvre eventually chosen is instructed using aural annunciations, and the vertical rate range required to obtain safe separation is indicated on the vertical speed indicator. When safe separation has been achieved, an explicit 'CLEAR OF CONFLICT' advisory is issued (FAA 2000; RTCA 1997).

Furthermore, TCAS is capable of coordinating conflict resolution with other TCAS-equipped aircraft. Since TCAS-equipped aircraft will, for the vast majority of cases, classify each other as threats at slightly different points in time, coordination is usually straightforward: the first aircraft selects and transmits its geometry-based sense and, slightly later, the second aircraft, taking into account the first aircraft's coordination message, selects and transmits the opposite sense. In certain special cases such as simultaneous coordination, both aircraft may select and transmit a sense independently based on the perceived encounter geometry. If these senses turn out to be incompatible, the aircraft with the higher Mode S address will reverse its sense, displaying and announcing the reversed RA to the pilot (RTCA 1997).

2.2.3 Flight Crew Interface

The flight crew interface of TCAS encompasses three major components, the traffic display, a control panel and means of presenting resolution advisories.

The TCAS traffic display is intended to aid pilots in the visual acquisition of traffic, including the differentiation of intruder threat levels, to provide situational awareness and to instil confidence in the presented resolution advisories.

Traffic is displayed in relation to a symbol reflecting ownship position, on a display featuring range rings, see Fig. 5. The symbology used to represent other traffic depends on the alert state and proximity of the corresponding traffic, as shown in Fig. 6.

Traffic symbols are supplemented by a two-digit altitude data tag in the same colour as the intruder symbol, indicating the altitude of the intruder aircraft in relation to ownship in hundreds of feet. For traffic above the own aircraft, the tag is placed above the traffic symbol and preceded by a '+', whereas other aircraft below ownship feature the tag below the symbol with a '−' as prefix. If other traffic is climbing or descending with a rate of 500 ft/min or more, an upward or downward arrow is displayed next to the symbol.

With a TA or RA in progress, the aircraft causing the alert and all proximate traffic within the selected display range is shown. However, it is recommended that all other traffic is displayed as well to increase the probability that pilots visually acquire the aircraft actually causing the alert.

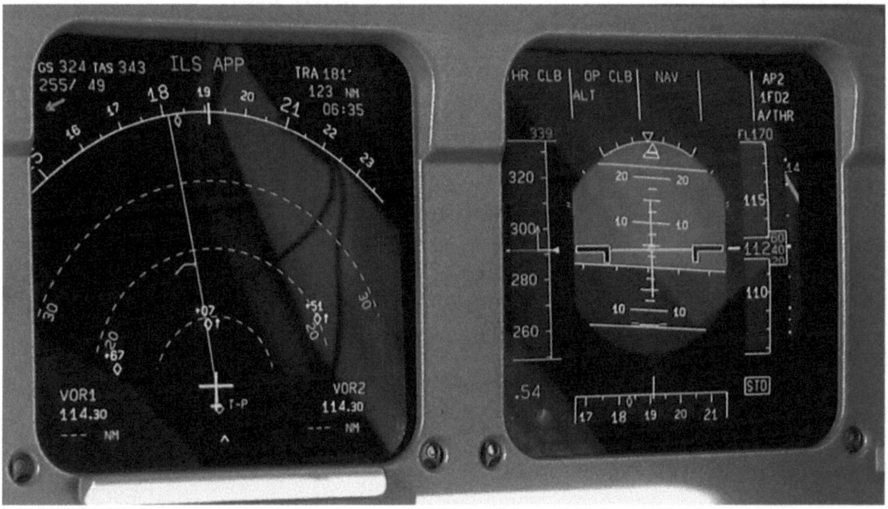

Fig. 5 Airbus A320 family EFIS with TCAS traffic presented on ND

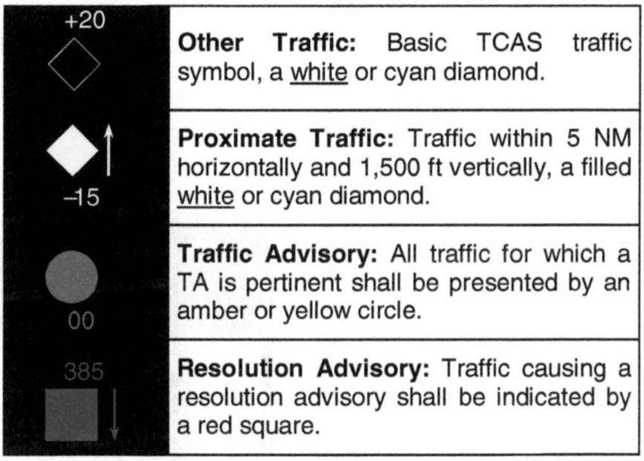

Fig. 6 TCAS symbology (according to RTCA (1997))

Table 1 details the aural annunciations for the different types of Resolution Advisories. In some implementations, aural annunciations are supplemented by textual messages (essentially repeating the callout) at a prominent location in the flight deck. In the presence of multiple RAs, the presentation of alerts to the flight crew is combined, such that the most demanding active RA is displayed. Figure 7 shows a TCAS vertical speed range indicated on the vertical speed scale of a PFD. The presentation details whether an RA requires corrective action, or merely warns against initiating an action that could lead to inadequate vertical separation. All positive RAs are displayed with a green "fly-to" indication detailing the corrective

Table 1 TCAS RA aural annunciations and visual alerts (RTCA 1997)

Resolution advisory type	Aural annunciation	Visual alert
Corrective climb	*Climb, Climb*	CLIMB
Corrective descend	*Descend, Descend*	DESCEND
Altitude crossing climb (corrective)	*Climb, Crossing Climb – Climb, Crossing Climb*	CROSSING CLIMB
Altitude crossing descend (corrective)	*Descend, Crossing Descend – Descend, Crossing Descend*	CROSSING DESCEND
Corrective reduce climb	*Adjust Vertical Speed, Adjust*	ADJUST V/S
Corrective reduce descent	*Adjust Vertical Speed, Adjust*	ADJUST V/S
Reversal to a climb (corrective)	*Climb, Climb NOW – Climb, Climb NOW*	CLIMB NOW
Reversal to a descend (corrective)	*Descend, Descend NOW – Descend, Descend NOW*	DESCEND NOW
Increase climb (corrective)	*Increase Climb, Increase Climb*	INCREASE CLIMB
Increase descent (corrective)	*Increase Descent, Increase Descent*	INCREASE DESCENT
Initial preventive RAs	*Monitor Vertical Speed*	MONITOR V/S
Non-crossing, maintain rate RAs (corrective)	*Maintain Vertical Speed, Maintain*	MAINTAIN V/S
Altitude crossing, maintain rate RAs (corrective)	*Maintain Vertical Speed, Crossing Maintain*	MAINTAIN V/S CROSSING
Weakening of corrective RAs	*Adjust Vertical Speed, Adjust*	ADJUST V/S
Clear of conflict	*Clear of Conflict*	CLEAR OF CONFLICT

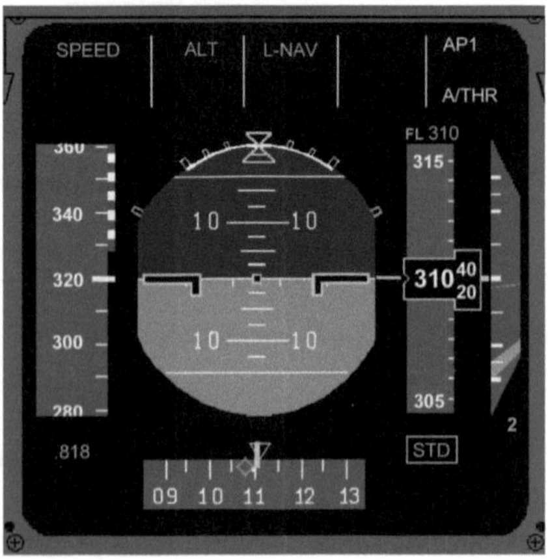

Fig. 7 TCAS vertical speed command indication on a PFD FAA (2000)

action. By contrast, vertical speed ranges to be avoided, including all negative RAs, are represented in red (RTCA 1997). In the example shown, the flight crew would have to (maintain) descent with 2,500 ft/min.

2.2.4 Automatic Dependent Surveillance: Broadcast (ADS-B)

Automatic Dependent Surveillance – Broadcast (ADS-B) is a function enabling aircraft, irrespective of whether they are airborne or on the ground, and other relevant airport surface traffic, to periodically transmit their state vector (horizontal and vertical position, horizontal and vertical velocity) and other surveillance information for use by other traffic in the vicinity or users on the ground, such as Air Traffic Control or the airline. This surveillance information, which is dependent on data obtained from various onboard data sources, such as the air data or navigation system, is broadcast automatically and does not require any pilot action or external stimulus. Likewise, the aircraft or vehicle originating the broadcast does not receive any feedback on whether other users within range of this broadcast, either aircraft or ground-based, decode and process the ADS-B surveillance information or not. ADS-B was conceived to support improved use of airspace, reduced ceiling and visibility restrictions, improved surface surveillance and enhanced safety such as conflict management (RTCA 2003). As an example, ADS-B may extend surveillance coverage and provide ATS services in airspace currently not served by radar-based surveillance systems, or support advanced concepts such as Free Flight in all airspace while maintaining a high level of flight safety (RTCA 2002).

There will be ADS-B mandates in Australia, Canada, Europe and the United States in the foreseeable future (Eurocontrol 2008; FAA 2007), since ADS-B is a core technology for the future ATM concepts currently under development.

In principle, ADS-B defines a set of data to be exchanged and is independent of any particular type of data link. The only requirement concerning the data link is that it must be capable of all-weather operations and fulfil the basic ADS-B capacity and integrity requirements. ADS-B system capacity was derived on the basis of 2020 traffic estimates for the Los Angeles basin. It is assumed that, for a high-density scenario, there will be approximately 2,700 other aircraft within 400 NM, of which 1,180 are within the core area of 225 NM, and 225 on the ground. The nominal ADS-B range is between 10...120 NM, depending on the type of emitter. ADS-B integrity, i. e. the probability of an undetected error in an ADS-B message although source data are correct, must be 10^{-6} or better on a per report basis (RTCA 2002).

The ADS-B implementation that will eventually become an international standard is based on the Mode S transponder. Apart from the normal, periodically transmitted Mode S squitters (DF = 11), which are, among others, used by multilateration installations and ACAS, there is an additional 112 bit Mode S Extended Squitter (ES) message (DF = 17), which is broadcast via the 1,090 MHz downlink frequency. Apart from the broadcasting aircraft's unique ICAO 24 bit address and a capability code, it contains a 56 bit message field (ME) that can be filled with one of the following sub-messages (RTCA 2003):

- *Airborne Position Message*: As soon as the aircraft is airborne, it automatically transmits its position, encoded via latitude/longitude, and altitude. Position information is supplemented by a Horizontal Containment Radius Limit (RC), expressed via the Navigation Integrity Category (NIC).

- *Surface Position Message*: While the aircraft is on the ground, it broadcasts this message containing a specifically encoded ground speed, heading or track, latitude/longitude and the Containment Radius (RC), instead of the Airborne Position Message.
- *Aircraft Identification and Type Message*: Irrespective of flight status, a message containing an 8-character field for the aircraft flight-plan callsign (e.g. an airline flight number) if available, or the aircraft registration, is broadcast. For surface vehicles, the radio call sign is used. The message also contains the ADS-B emitter category, permitting a distinction between aircraft (including different wake vortex categories), gliders/balloons and vehicles, where e.g. service and emergency vehicles can be distinguished.
- *Airborne Velocity Message*: While airborne, ADS-B equipped aircraft transmit their vertical and horizontal velocity, including information on the Navigation Accuracy Category (NACV).
- *Target State and Status*: This message conveys an aircraft's target heading/track, speed, vertical mode and altitude. In addition, it contains a flag detailing ACAS operational status, and an emergency flag that can be used to announce presence and type of an emergency (i.e. medical, minimum fuel, or unlawful interference). However, the message is only transmitted when the aircraft is airborne, and if target status information is available and valid.
- *Aircraft Operational Status*: This message provides an exhaustive overview of an aircraft's operational status, including both equipment aspects (such as the availability of a CDTI) and navigation performance, e.g. the Navigation Integrity Category (NIC), the Navigation Accuracy Category (NAC) and the Surveillance Integrity Level (SIL). This is intended to enable surveillance applications to determine whether the reported position etc. has an acceptable accuracy and integrity for the intended function.

In addition, besides status and test messages, several further formats are envisaged for future growth potential such as the exchange of trajectory data. Non-transponder broadcasters such as airport vehicles use a different downlink format (DF = 18), and DF = 19 is reserved for military aircraft and applications.

The basic ADS-B application from an onboard perspective is a Cockpit Display of Traffic Information (CDTI), which is mainly intended to improve awareness of proximate traffic, both in the air and on the ground. In addition, depending on how complete the coverage of surrounding traffic is, a CDTI may also serve as an aid to visual acquisition, thus supporting normal "see and avoid" operations. A CDTI is not limited to a single or particular source of traffic data. Therefore, information from TIS-B or TCAS can be integrated as well. It is important to note that a CDTI is conceptually not limited to traffic presentation, but can also be used to visualise weather, terrain, airspace structure, obstructions, detailed airport maps or any other static information deemed relevant for the intended function. Last but not least, a CDTI is required for a number of envisaged future onboard applications (see below), particularly Free Flight operations (RTCA 2002).

ADS-B is also seen as a valuable technology to enhance operation of Airborne Collision Avoidance Systems (ACAS), e.g. by increased surveillance performance, more accurate trajectory prediction, and improved collision avoidance logic. In particular, the availability of intent information in ADS-B is seen as an important means of reducing the number of unnecessary alerts. Furthermore, DO-242A proposes extended collision avoidance below 1,000 ft above ground level and the ability to detect Runway Incursions as potential ADS-B applications and benefits (RTCA 2002).

Aircraft conflict management functions are envisaged to be used in support of cooperative separation whenever responsibility has been delegated to the aircraft. Airspace deconfliction based on the exchange of intent information will be used for strategic separation. Since pilot and controller share the same surveillance picture, resolution manoeuvres are expected to be better coordinated (RTCA 2002).

Other potential ADS-B applications mentioned in DO-242 A include general aviation operations control, improved search and rescue, enhanced flight following, and additional functions addressing Aircraft Rescue and Fire Fighting (ARFF) or other airport ground vehicle operational needs (RTCA 2002).

2.2.5 Traffic Information Service: Broadcast (TIS-B)

The fundamental idea behind TIS-B is to extend the use of traffic surveillance data readily available on the ground, which are typically derived from ATC surveillance radars, multi-lateration installations or other ground-based equipment, to enhanced traffic situational awareness and surveillance applications aboard aircraft or airport vehicles. Essentially, TIS-B is therefore a function in which transmitters on the ground provide aircraft, airport vehicles or other users with information about nearby traffic (RTCA 2003).

When based on ATC surveillance data, TIS-B enables pilots, vehicle drivers and controllers to share the same traffic picture, which is important in the context of ATM concepts promulgating collaborative decision making. Likewise, TIS-B addresses the essential problem of non-cooperative traffic, and thus the issues concerning the completeness and quality of the traffic surveillance picture available on board, particularly in a transition phase with ADS-B equipage levels still low. In fact, TIS-B is sometimes only regarded as an interim or back-up solution (RTCA 2003).

Apart from ground-based surveillance sensors, TIS-B information might also be based on received ADS-B Messages originally transmitted on a different data link (RTCA 2003); from this perspective, TIS-B would include ADS-R. For the purpose of this thesis, however, these services are considered separately, under the notion that ADS-R is a simple repeater function, whereas TIS-B typically involves sophisticated sensor data fusion.

Originally, like ADS-B, TIS-B was a data exchange concept not specifying a particular data link or exchange protocol. However, as for ADS-B, the use of the Mode S data link capability is defined in RTCA DO-260A (RTCA 2003).

Technically, the main difference between ADS-B and TIS-B is that the data transmitted via TIS-B are acquired by ground-based surveillance such as radar and may thus include aircraft not equipped with ADS-B or even a Mode S-transponder.

Each TIS-B message contains information on a single aircraft or vehicle, i.e. the Mode S message load increases dramatically in high-density airspace. Another concern related to TIS-B is that its update rate will typically be significantly lower than the ADS-B update rate, because secondary surveillance radars, depending on their rotation speed, only provide updates every 5–12 s. Additionally, it is believed that latency effects will limit the operational usability of TIS-B data.

2.3 Controller-Pilot Data Link Communications (CPDLC) and Future Air Navigation System (FANS)

Concerning the interaction with Air Traffic Services, CPDLC and Automatic Dependent Surveillance (ADS) first emerged in an operational context over the South Pacific Ocean (Billings 1996), an area inaccessible to the means of ATC commonly employed in continental airspace. Previously, the absence of radar surveillance and the fact that aircraft operate beyond the range of VHF stations required procedural ATC using HF radio, which only has very poor voice quality due to fading, and relying on crew reports at certain pre-defined waypoints. This resulted in an inflexible system necessitating very large safety distances between adjacent aircraft.

By contrast, CPDLC permits to sustain data link communications between pilot and controller, particularly in areas where voice communications are difficult. Like conventional R/T, CPDLC is based on the exchange of standardised message elements, which can be combined to complex messages, if necessary. The philosophy that the crew is obliged to read back all safety-related ATC clearances is retained in a CPDLC environment (ICAO 2001a; 2002) by the implementation of a dedicated acknowledgement process (FANS 2006). The approximately 180 pre-formatted CPDLC message elements defined for the FANS A (via ACARS) and B (ATN) implementations are analogues of the existing ICAO phraseology, with the important difference that CPDLC messages are machine-readable. This is achieved by allocating a certain identification code to each individual message element, and permitting certain parameters such as flight level, heading or waypoint names to be transmitted separately as variables (FANS 2006).

Until today, the main application of CPDLC is in oceanic airspace. Besides, it is widely used in conveying pre-departure clearances to aircraft at airports to avoid radio frequency congestion and to reduce the chances of errors in the clearance delivery process. It is important to note that operational use is currently limited to fields where the interaction between pilot and controller is not time critical.

In Europe, CPDLC is currently envisaged as a supplementary means of communication and not intended to replace voice as the primary means of communication.

Fig. 8 Location of DCDU equipment in an Airbus A330 cockpit

Rather, it will be used for an exchange of routine, non-time critical instructions, clearances and requests between flight crew and controller (Fig. 8).

Initially, it will be applied to upper airspace control above FL 285 (Eurocontrol 2009). Accordingly, initial trials using the Aeronautical Telecommunication Network (ATN) are taking place at Maastricht UAC. From 2013 to 2015, the ATM-side introduction will take place. For aircraft, a forward fit of new airframes is foreseen from 2011 onwards and retrofit by 2015. In any case, the operational decision to use CPDLC rests with the flight crew and the controller.

Figure 9 shows a typical FANS B architecture. Aboard the aircraft, a so-called Air Traffic Services Unit (ATSU) manages communication with AOC and ATC via the various available data links, some of which are customer options. The main difference compared to FANS A is that communication with ATC occurs via the ATN instead of ACARS. In the cockpit, the flight crew is notified of a new incoming ATC message by an indicator on the glareshield and a ringing sound. Pilots may then review the ATC instruction and acknowledge it, reject it or ask for more time to assess its impact. On the Airbus A320/A330/A340 aircraft families, the so-called Datalink Control and Display Unit (DCDU), illustrated in Fig. 8, is used to review and acknowledge ATC CDPDLC messages. In case the flight crew needs to initialise a request to ATC, the Multipurpose Control Display Unit (MCDU) is used to compose the CPDLC message, which is the forwarded to the DCDU for review prior to transmission (de Cuendias 2008; FANS 2006).

Advantages and drawbacks of CPDLC have been discussed at length for some years. CPDLC reduces misunderstandings and errors pertaining to poor voice quality, fading or language issues, and ensures that flight crews do not erroneously react to instructions intended for another aircraft (e.g. due to similar callsigns).

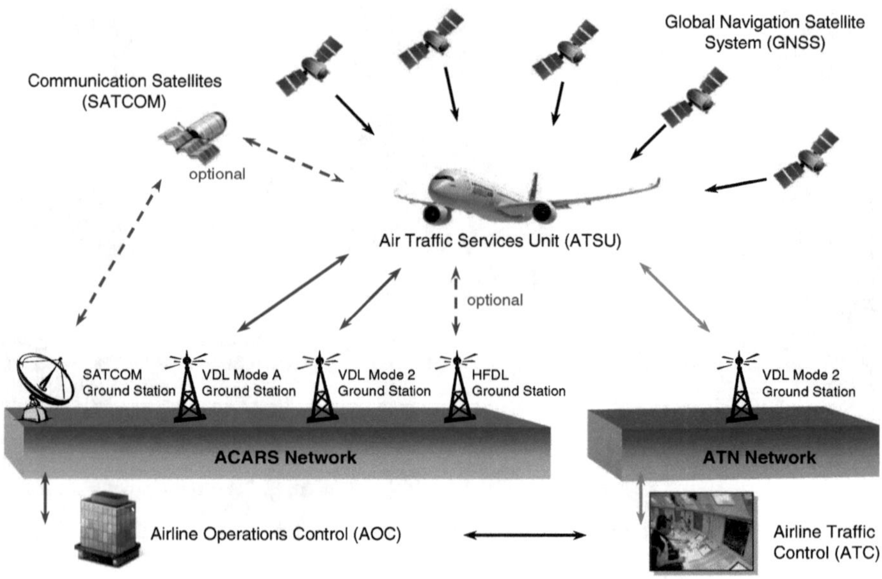

Fig. 9 FANS B architecture for CPDLC and AOC communication (according to de Cuendias (2008))

Furthermore, the possibility of recalling messages makes ATC clearances and instructions continuously accessible, and transcription errors are reduced by the possibility of transferring CPDLC information directly to the FMS. On the other hand, particularly the preparation of a request to ATC takes much longer with CPDLC, and any party-line information is lost. Last but not least, the introduction of CPDLC transfers additional information to the already heavily loaded visual channel (Airbus 2004a).

2.4 Data Link Automatic Terminal Information Service (D-ATIS)

According to ICAO Annex 11, Automatic Terminal Information Service (ATIS) has to be provided at aerodromes whenever there is a need to reduce frequency congestion and air traffic controller workload in the terminal area. ATIS contains information pertaining to operations relevant to all aircraft such as meteorological conditions, runway(s) in use and further information which may affect the departure, approach and landing phases of flight, thus eliminating the need of transmitting the corresponding items individually to each aircraft as part of ATC communication. ATIS messages are updated at fixed intervals (usually 30–60 min) and whenever a change of meteorological or operational conditions results in substantial alterations to their content. In order to make consecutive ATIS transmissions distinguishable and to enable controllers to confirm with flight crews that they have

Fig. 10 Sample D-ATIS for Bremen airport

```
.D-ABET    ----    DLH9YV        28JUL06  0735Z
ATIS BRE
ATIS E RWY: 27 TRL: 70 SR: 0336 SS: 1927
METAR 280720 EDDW 20004KT 7000
NSC 21 /20
1012
TEMPO 4000 TSRA BKN030CB
SUSAN EXPECT ILS APCH ATTN ATTN NEW DEP
FREQ 1 2 4 DECIMAL 8
CONTRACT ACTIVE
```

the most recent information available, each ATIS broadcast is identified by a single letter of the ICAO spelling alphabet, which is assigned consecutively in alphabetical order (ICAO 1999, 2001).

Conventional ATIS transmits a voice recording typically lasting less than 1 min over a published VHF frequency in the vicinity of the aerodrome (ICAO 1999). To ensure that broadcasts do not exceed 30 s wherever practicable, larger airports offer different ATIS services for departing and arriving aircraft (ICAO 2001).

To improve ATIS reception by eliminating the need for flight crews to transcribe voice recordings, D-ATIS provides a textual version of the ATIS audio broadcast. Content and format of D-ATIS and ATIS messages should be identical wherever these two services are provided simultaneously, and both services should be updated at the same time (Fig. 10).

D-ATIS is already available via ACARS at many airports. It is expected, however, that this service will migrate to VDL mode 2. A problem associated with D-ATIS is that the message content is not standardised sufficiently enough to make it machine-readable. An advantage of D-ATIS, however, is that it is available independently of the distance from the airport, while conventional ATIS is only available within the range of the airport VHF transmitter. While this limitation is marginal in routine operations, it will become relevant when operational conditions mandate a diversion. This may be of particular relevance for aircraft with specific constraints with respect to runways, such as the Airbus A380. If, for example, an airport has only one runway supporting the A380 and this runway is closed down while the aircraft is already in flight, the flight crew could be informed of this situation via D-ATIS as soon as it occurs, whereas they would probably not be notified before in range of the airport VHF transmitter otherwise. Thus, D-ATIS increases the time for strategic re-planning of the flight.

2.5 Synthetic Vision for Improved Integrated Information Presentation

Collisions with terrain, called "Controlled Flight into Terrain" (CFIT), are still among the main causes for fatal airliner accidents (Boeing 2004). This type of accident is characterised by the fact that pilots inadvertently fly their fully functional and controllable aircraft into terrain because they misjudge their position.

Fig. 11 PFD with basic terrain colouring and intuitive terrain warning

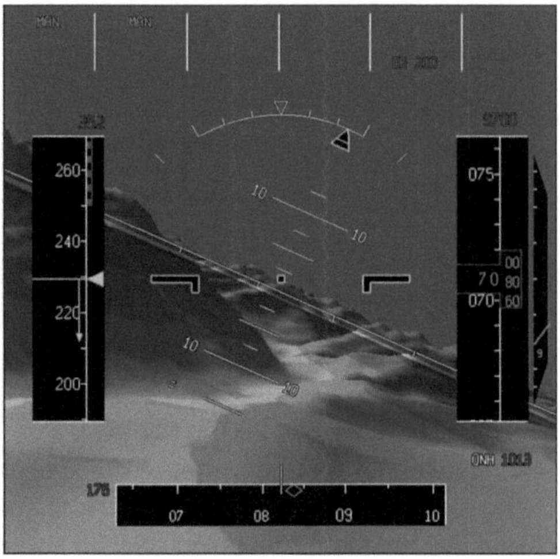

To prevent such accidents, regulators require that all aircraft used in scheduled airline operations have to be equipped with a Terrain Awareness and Warning System (TAWS). Since most CFIT accidents occur in less favourable visibility conditions, a supplement to this warning system approach consists of a so-called "Synthetic Vision System" (SVS), a database-driven representation of terrain and airport features resembling the real outside world. Typically, the black background of the conventional Electronic Flight Instrument System (EFIS) is replaced by a three-dimensional perspective, database-generated synthetic view of the flight environment ahead of the aircraft on the Primary Flight Display (PFD). In enhancement of the artificial horizon concept, terrain information on the PFD is thus presented based on the pilots' frame of reference, similar to the real world outside, see Fig. 11. As a consequence, the synthetic view on the display matches the view the crew would have when looking out of the front cockpit window in perfect visibility. This generic terrain representation, which can also be added to map-type displays using an orthographic projection, is intended to enable the crew to judge their position in relation to terrain, regardless of visibility restrictions due to meteorological or night conditions.

Therefore, the ultimate goal of SVS terrain representation is to achieve a substantial improvement of the pilot's situational awareness with respect to terrain, and it is also envisaged to support pilots with intuitive predictive alert cues. For this purpose, terrain above the own aircraft's altitude can be colour-coded, thus aiding flight crews in maintaining a route that will stay clear of terrain conflicts at all times. Unintentional deviations from chart-published routes will become apparent in an intuitive fashion. Such an SVS functionality is expected to be especially beneficial in the event of unplanned deviations from the original flight plan, e.g. during an emergency descent over mountainous terrain.

A common misunderstanding about Synthetic Vision Systems is that they violate the so-called "dark and silent cockpit" philosophy, which permits to display only

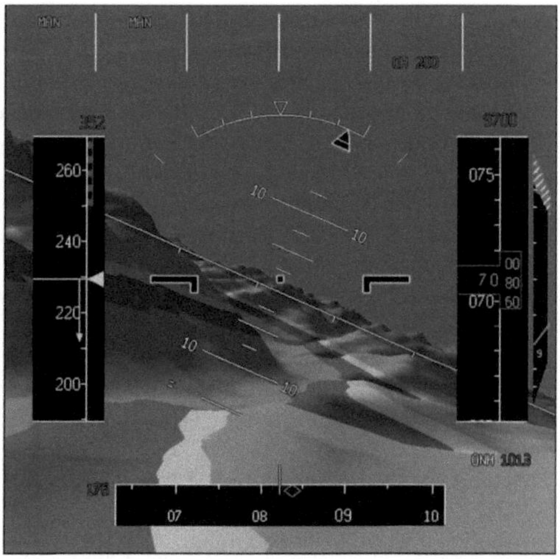

Fig. 12 PFD with terrain hazard display colouring generated by the TAWS

information which is relevant for the pilot to complete his or her current task. However, this relates to abstract 2D symbology and alphanumerical information requiring active analysis and interpretation by the crew. In contrast to this, the presentation of terrain information corresponds to natural human perception. It is thus very important to differentiate information that has to be processed consciously from information which can be perceived intuitively (Vernaleken et al. 2005).

To provide a vertical reference, the basic Gouraud-shaded terrain representation can be superimposed with a layering concept based on a variation of the brightness of the basic terrain colour (see Fig. 12) or the TAWS terrain colours (cf. Fig. 12).

In this context, it has to be stressed that a photorealistic representation, e.g. with satellite images mapped as textures on the digital terrain model, is neither required nor desirable for an SVS system. Photorealism not only limits the possibilities of augmenting the SVS with artificial cues such as the intuitive terrain warning, but also adds information on the nature of vegetation or the location of small settlements, which is hardly relevant for the piloting task while simultaneously creating new issues. As an example, the way of reflecting seasonal changes is an unsolved problem and usually neglected in studies promoting photorealistic terrain representation. In conclusion, if considered useful for orientation, cultural information on rivers, highways and metropolitan areas can be conveyed more efficiently in the form of vector data overlaying the digital terrain elevation model (Vernaleken et al. 2005).

References

Airbus, S. A. S. (2005). *Airbus A320 flight crew operating manual* (Vol. 1). Systems Description, Revision 37, Toulouse-Blagnac.

Airbus Customer Services. (2004a). *Getting to grips with datalink, flight operations support & line assistance*. Toulouse-Blagnac.

Airbus Customer Services. (2004b). *Getting to grips with FANS, flight operations support & line assistance*. Toulouse-Blagnac.
Billings, C. E. (1996). *Human-centered aviation automation: Principles and guidelines*, NASA Technical Memorandum No. 110381, Moffett Field.
Bureau d'Enquêtes et d'Analyses (BEA). (2009). *Interim report on the accident on 1st June 2009 to the Airbus A330-203, registered F-GZCP, operated by Air France*, Flight AF 447 Rio de Janeiro – Paris, Paris.
De Cuendias, S. (2008). *The future air navigation system FANS B: Air traffic communications enhancement for the A320 Family*. Airbus FAST Magazine No. 44, Tolouse-Blagnac.
Eurocontrol. (2009). *Flight crew data link operational guidance for LINK 2000+ Services*, Edition 4.0, Brussels.
Eurocontrol Safety Regulation Commission. (2008). *Annual safety report 2008*. Brussels.
FANS Interoperability Team. (2006). *FANS-1/A operations manual*, Version 4.0.
Federal Aviation Administration. (2007). *ADS-B aviation rulemaking committee*, Order 1110.147. Washington, DC.
Federal Aviation Administration (FAA). (2000). *Introduction to TCAS II, Version 7*. Washington, DC.
International Civil Aviation Organization (ICAO). (1999). *Manual of air traffic services data link applications*, Doc 9694, AN/955, Montreal: International Civil Aviation Organization.
International Civil Aviation Organization (ICAO). (2001). Annex 11 to the Convention on International Civil Aviation, Aircraft Traffic Services, Montreal.
International Civil Aviation Organization (ICAO). (2001a). *Air traffic management* (14th ed.), Doc 4444, ATM/501. Montreal: International Civil Aviation Organization.
International Civil Aviation Organization (ICAO). (2001b) *Annex 6 to the convention on international civil aviation, Operation of Aircraft, Part I: International commercial air transport – aeroplanes* (8th ed.). Montreal: International Civil Aviation Organization.
International Civil Aviation Organization (ICAO). (2002). *Annex 10 to the convention on international civil aviation, Aeronautical telecommunications (Vol. IV): Surveillance radar and collision avoidance systems* (3rd ed.). Montreal: International Civil Aviation Organization.
International Civil Aviation Organization (ICAO). (2004). *Annex 15 to the convention on international civil aviation*. Montreal: Aeronautical Information Services.
International Council on Systems Engineering (INCOSE). (2010). *INCOSE systems engineering handbook* (Vol. 3.2), INCOSE-TP-2003-002-03.2, San Diego.
Radio Technical Commission for Aeronautics (RTCA). (1997). DO-185A, *Minimum operational performance standards for traffic alert and collision avoidance system II (TCAS II) airborne equipment*. Washington, DC.
Radio Technical Commission for Aeronautics (RTCA). (2002). DO-242A, *Minimum aviation system performance standards for Automatic Dependent Surveillance Broadcast (ADS-B)*. Washington, DC.
Radio Technical Commission for Aeronautics (RTCA). (2003). DO-260A, *Minimum operational performance standards for 1090 MHz extended squitter Automatic Dependent Surveillance – Broadcast (ADS-B) and Traffic Information Service – Broadcast (TIS-B)*. Washington, DC.
The Boeing Company. (2004). *Statistical summary of commercial jet airplane accidents*. Worldwide Operations 1959–2003. Seattle.
Vernaleken, C. (2011). *Autonomous and air-ground cooperative onboard systems for surface movement incident prevention*. Dissertation, TU Darmstadt.
Vernaleken, C., von Eckartsberg, A., Mihalic, L., Jirsch M., & Langer, B. (2005). The European research project ISAWARE II: A more intuitive flight deck for future airliners. In *Proceedings of SPIE, enhanced and synthetic vision* (Vol. 5802). Orlando.

Military Aviation

Robert Hierl, Harald Neujahr, and Peter Sandl

1 Introduction

In today's military aviation, information management is the crucial factor for mission success. Proper decision-making depends not only on the amount and quality of data, but to the same degree on the consideration of its context and the combination with other data. Without the appropriate information the pilot of a combat or surveillance aircraft or the commander of an unmanned aerial vehicle (UAV) is unable to adequately fulfil his mission.

Not only the political situation but also technology has changed drastically since the Cold War. Many of the current military conflicts are of asymmetric character; a *classical* battlefield often does not exist. In many cases military actions take place in an environment where the combatants are difficult to distinguish from the civil population. The opponents in such conflicts cannot be easily identified – they often don't wear a uniform. Another aspect is that in recent years, military conflicts with involvement of western military forces are at the focus of public interest. The appropriateness of military decisions is critically assessed and the acceptance of both civil victims and own losses is decreasing. Therefore, present-day rules of engagement stipulate the protection of non-involved people, which in turn means that military operations have to be performed with very high precision. This is only possible if the protagonists receive the necessary information in real-time.

The vast amount of information of variable quality that is available today, nearly everywhere in the world and in near real time, is often associated with the expression *Network Centric Warfare* (cf. Alberts et al. 2001). It implies that military success nowadays not only relies on the qualities of a particular weapon

R. Hierl (✉)
Flight Test, Technical Centre of Aircraft and Airworthiness WTD61, Manching, Germany
e-mail: RobertHierl@bwb.org

H. Neujahr • P. Sandl
Human Factors Engineering, Cassidian, Manching, Germany

M. Stein and P. Sandl (eds.), *Information Ergonomics*,
DOI 10.1007/978-3-642-25841-1_6,

system like a fighter aircraft, but that primarily the connection of information from multiple systems, expressed by the term *System of Systems*, allows for true military superiority. This has profound implications on the design of the human–machine interfaces of basically any military system. Today combat aircraft or UAV ground stations are integral parts of this information network, within which the pilot or the UAV operator acts as an executive unit that has any freedom with respect to local or tactical decisions. In order to perform his task under the given conditions, it is essential that the information provided to the operator is up-to-date, consistent and complete, and that it is appropriately presented.

1.1 Mission Diversity

Military aviation is characterised by a high diversity of missions. Traditionally, each of these missions had to be fulfilled by a particular type of aircraft. Already during the First World War the air forces distinguished between fighters, bombers, fighter-bombers and reconnaissance planes.

The basic task of an aircraft is called a *role*. Military heavy aircraft for example may have a strategic and tactical air transport role, a bombing role, an aerial refuelling role, a maritime patrol role or a C^3 (Command, Control and Communications) role, to name just a few. The roles of the more agile fighter aircraft are air-to-air combat, air-to-surface attack or, since the Vietnam War, electronic warfare.[1] Later, with the introduction of the fourth generation of combat aircraft (examples see below) the paradigms changed: instead of requiring dedicated aircraft for specific roles, e.g. fighters, bombers, reconnaissance aircraft, the forces now asked for a multi-role capability. The root cause of this new philosophy was a dramatic rise in development costs since the early 1950s. Simply speaking, different aircraft for different roles were no longer affordable. To enhance effectiveness and economy, it was now required that a single type of aircraft was able to fulfil different roles during the same mission. The idea was that a small group of swing role capable aircraft should be able to achieve the same military goals as a much larger package of fighter bombers and escort fighters which would traditionally have been employed for that task. By providing their own protection, i.e. acting as fighters and bombers simultaneously, the new airplanes would render a dedicated escort redundant. All European fourth generation combat aircraft such as the Eurofighter *Typhoon* (*Swing-Role*[2]), the Dassault *Rafale* (*Omni-Role*[3]) and the Saab *Gripen* (*JAS*,[4] i.e. "Jakt, Attack, Spanning": Fighter, Bomber, Recce)

---

[1] For a definition see Joint Chiefs of Staff (2010), p. 179.

[2] http://www.eurofighter.com/eurofighter-typhoon/swing-role/air-superiority/overview1.html; Accessed 28 April 2011.

[3] http://www.dassault-aviation.com/en/defense/rafale/omnirole-by-design.html?L=1; Accessed 28 April 2011.

[4] http://www.saabgroup.com/en/Air/Gripen-Fighter-System/Gripen/Gripen/Features/; Accessed 28 April 2011.

followed this trend. Even the American top fighter Lockheed-Martin F-22 *Raptor*, originally designed as a pure air-to-air aeroplane, was given a secondary air-to-surface capability. UAVs are designed to fulfil predominantly surveillance or reconnaissance roles, but there are also unmanned systems with a delivery capability for effectors like the General Atomics *Predator/Reaper*.

1.2 How the Tasks Determine the Design of the HMI

Information systems in modern military aircraft provide the pilot or the UAV commander with the information he needs to fulfil his mission. To better understand the requirements for the human machine interface of modern fighters, a historical review is helpful.

The tasks of an aircraft pilot can be structured as follows (Billings 1997, p. 16):

- Aviate, i.e. control the airplane's path,
- Navigate, i.e. direct the airplane from its origin to its destination,
- Communicate, i.e. provide, request, and receive instructions and information, and, particularly and increasingly in modern aircraft,
- Manage the available resources.

Another task is specific for military operations. This is

- The management of the mission, i.e. the fulfilment of a military task as defined by the aircraft role in a tactical environment.

To **aviate** means to control the path of the aircraft in a safe manner. This was the sole and challenging task of the pilot in the pioneering days of aviation. The necessary information with the exception of the airspeed was obtained from outside references and by the haptic perception of the aircraft's accelerations. Pilots call this the "seat of the pants" cues. At that time the pilots flew their planes mostly by feel rather than by reference to instruments. Although these basic cockpit instruments became more refined in the course of time, cockpit design as such was not an issue yet. This was the state of the art during the First World War.

A big step was achieved when instrument flying, or *blind flying*, as it was called at that time, was introduced in the 1920s. To enable safe, controlled flight in clouds, the pilot needed a reference, which enabled him to keep his aircraft straight or perform shallow turns while no visual reference was available. In clouds, the *seat of the pants* proved to be no longer helpful, on the contrary, the sense of balance in the inner ear could easily mislead the human operator; this state became known as spatial disorientation. Deadly spins, structural failure due to excess speed or flight into terrain occurred when pilots flew in clouds for a prolonged time without blind flying instruments. Eventually, slip ball and turn needle were introduced as simple instrument flying aids and were added to the standard set of cockpit gauges.

The **navigate** task was introduced when the intended purpose was to fly to a particular destination. Under good weather conditions, it was possible to stick to visual references and continuously follow landmarks. However, in poor visibility or over the sea it was necessary to determine the correct course over ground by means of a new cockpit instrument: the compass. Airspeed and travel time allowed an estimation of the covered distance. Modern navigation systems are based on a combination of Inertial Reference Systems and Satellite Navigation (GPS), and the pilot obtains extremely precise three-dimensional steering information from an integrated Flight Management System.

With regard to the **communication** task it is common to distinguish between aircraft internal and external communication. In the early days, pilots flew alone, an aviation infrastructure did not exist and hence there were neither a cause nor the technical facilities to communicate. Internal communication came into play when increased aircraft complexity made additional crew members necessary.

The first kind of external communication was the exchange of information between ground crew and pilot by means of semaphores and hand signals. Later, when the aeroplanes began to be co-ordinated from the ground by military control centres or air traffic control the communication was executed verbally via radios. The pilots had to know to whom to speak, who was the responsible point of contact, and what was the correct transmission frequency.

To cope with the communication requirements of today's air traffic and battle-field management, modern aircraft feature complex communication systems with several radios and data link capability.

With the introduction of increasingly complex aircraft systems such as the Flight Control System, the Guidance and Navigation System, the Communication System and, in parallel, the improved handling qualities of modern aircraft, a significant shift away from the basic *aviate* task of the early days towards the **manage**ment of systems and information was the logical consequence. The military mission has always posed additional demands on the pilot. Traditionally, military aviators are briefed before their missions. In the first half of the twentieth century this was the information which they based upon their decisions. Due to the absence of on-board sensors and information networks, there was little extra information they had to cope with while they executed a flight. This started to change in the beginning of the 1970s. The introduction and continuous improvement of a multitude of on-board sensors, for example the radar, as well as the emergence of the data link technology, which supplemented voice communication by data communication, caused an evolution, which can be characterised as an "information explosion". The consequence was a drastic change of the pilot's tasks: rather than concentrating on aircraft control, he is now focussed on managing the available information. The progress in a number of technical areas, predominantly microelectronics and display technology was a necessary prerequisite. Through appropriate automation measures it became possible to simplify the original piloting task such that cognitive capacities were made available for the more sophisticated mission management tasks. Appropriate measures in the HMI design were taken to make sure that the resulting workload of the pilot is kept within acceptable limits.

1.3 Human Factors in Aviation

A reflection on the chronological cockpit development since the beginning of aviation shows that the number of cockpit instruments increased continuously until around 1970 (see Fig. 1). Thereafter a trend reversal can be observed, which, according to Sexton (1988), is attributed to the introduction of multifunction displays and digital avionics. The diagram in Fig. 1 corresponds to civil aircraft. For the military sector, however, the trend is very similar, except that the peak of the curve is a few years earlier. The first fighter aircraft which featured electronic displays, at that time monochrome Cathode Ray Tubes (CRTs), and a Head-Up Display but also traditional instruments, for example the McDonnell Douglas F-15 *Eagle* or the Panavia *Tornado*, had their first flights in the mid-1970s, while the first airliners with glass cockpits (see Sect. 2.2 below), e.g. the Airbus A310 or the Boeing B757 had their maiden flights in the early 1980s.

The other aspect that determines the design of aircraft cockpits down to the present day is the increasing consideration of human factors. Before the Second World War there was virtually no systematic approach in human engineering, particularly with respect to cognitive aspects. Shortfalls in the interaction of the pilot with the aircraft were tried to be overcome by intensifying pilot training. The quest for tactical advantages in military air operations through ever increasing flight envelopes initiated a high effort in investigating flight medical aspects, aiming to protect the air crew from a hostile environment which is characterised by a lack of oxygen, low air-pressure and low temperatures. Furthermore, the improving

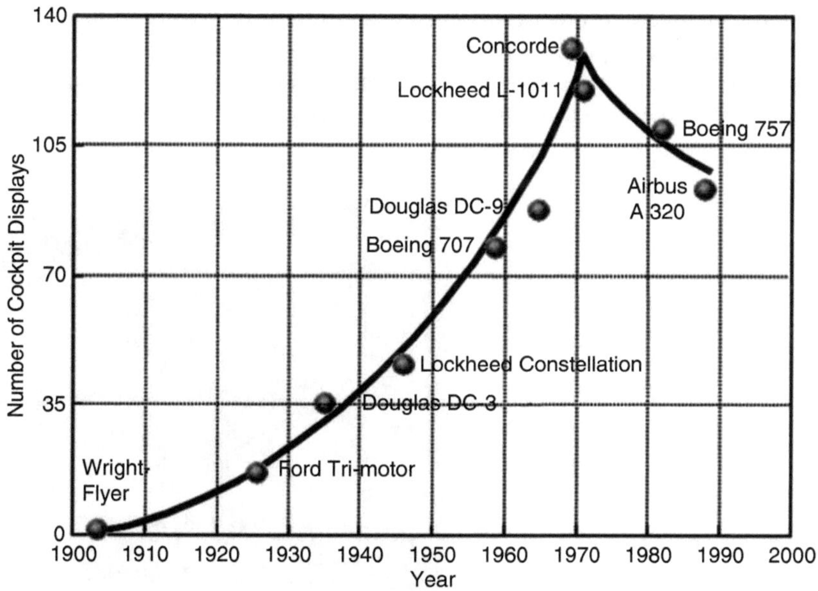

Fig. 1 Growth in number of displays (acc. to Sexton 1988)

manoeuvrability of the fighter aircraft resulted in the development of equipment for the protection of the pilot against high accelerations.

The clarification of mysterious accidents, which happened at the beginning of the 1940s with the aircraft types P-47, B-17, and B-25, is presumed to be one of the major milestones in aviation psychology. Roscoe (1997) wrote "Alphonse Chapanis was called on to figure out, why pilots and copilots ... frequently retracted the wheels instead of the flaps after landing. ... He immediately noticed that the side-by-side wheel and flap controls ... could easily be confused. He also noted that the corresponding controls on the C-47 were not adjacent and their methods of actuation were quite different; hence C-47 copilots never pulled up the wheels after landing." The problem was fixed by "a small, rubber-tired wheel, which was attached to the end of the landing gear control and a small wedge-shaped end at the flap control on several types of airplanes, and the pilots and copilots of the modified planes stopped retracting their wheels after landing" (Roscoe 1997). This shows that the erroneous operation had to be attributed rather to an inappropriate information presentation, which is a cockpit design error (Roscoe 1997), than to inattentive pilots.

After the Second World War the so-called Basic T or Cockpit-T became the standard for cockpit instrumentation (cf. Newman and Ercoline 2006). It represents a consistent array of the most important gauges: air speed indicator, attitude indicator, altitude indicator and compass.

Figure 2 shows an example of such an arrangement even highlighted by a white outline around the gauges on the main instrument panel. The other instruments were placed without following a particular standard. The T-shaped has two advantages: firstly, it provides the quickest and most effective scan of the most relevant information, which is particularly essential during instrument flights, where pilots have to adhere to precise parameters and refer exclusively to their cockpit gauges. Secondly, it supports type-conversion, since a pilot trained on one aircraft can quickly become accustomed to any other aircraft if the instruments are identical. Since the 1960s the T-shaped arrangement can be found in virtually

Fig. 2 Cockpit-T instrument arrangement. Source: http://commons.wikimedia.org/wiki/File:Trident_P2.JPG

all aircraft worldwide. The array of information provided by the conventional Cockpit-T today is even maintained in all aeroplanes equipped with a glass cockpit: airliners, fighter aircraft, but also, general aviation planes. Only the way the information is presented has changed towards a more display-compatible form (cf. Jukes 2003).

This article illustrates the role of information in modern military aviation. By means of examples from fighter aircraft and UAV operation, it describes how the designers of a cockpit or a UAV ground station are handling the amount of information available to make sure that the pilot or the UAV commander are able to fulfil their tasks in the most efficient manner.

2 Challenges for the Design of Modern Fighter Aircraft Cockpits

2.1 Introduction

As pointed out above, today's fighter aircraft might be regarded as complex information systems. They are participants in other, more extensive military networks with, which the pilot continually exchanges information. Apart from that there are a variety of other on-board information sources that have to be taken into account. To fulfil his complex mission and survive in a hostile environment, the military pilot must have the best and most up-to-date situational awareness (SA – an explanation of this term is given in the Cockpit Moding section below). SA forms the basis for his tactical decisions and is built up by means of the information available in and around the cockpit. Hence, today's fighter aircraft are in fact complex information systems with the capability to deploy weapons.

The design of a military fighter aircraft is based on a set of high-level requirements which in turn are derived from the intended operational role. Before the actual design work starts, a detailed analysis is carried out, during which the role of the weapon system under consideration is subdivided into different missions. Some of them are found to be design driving and are pursued. Each of these missions is then broken down into subtasks, e.g. Take-off, Climb, Manoeuvre at Supersonic Speed. Usually, performance requirements are defined for each subtask. The final design has to make sure that the levels of performance specified are attained. At the same time it must be assured that the flying qualities of the aircraft enable the pilot to conduct his mission safely and at an acceptable workload level. The pilot's workstation forms the HMI (Human–Machine Interface), consisting of displays and controls for interaction with the aircraft and its systems.

A big challenge for the design originates from the multirole capabilities (see previous section) required by many aircraft since more and more functions have to be integrated into a single platform.

The following sections address specific topics of a fighter cockpit with special focus on its information-mediating role. They cover general aspects of cockpit design illustrated by examples taken from the Eurofighter *Typhoon,* which features a modern state-of-the-art cockpit.

2.2 Cockpit Geometry and Layout

The shape of an aircraft is primarily determined by performance requirements such as maximum speed, manoeuvrability, range and endurance. It should also be mentioned that low observability, i.e. stealth, has become a very important, sometimes even the primary design goal, for example for the Lockheed F-117 *Nighthawk* or the Lockheed-Martin F-22 *Raptor*. The underlying physics of radar energy dictate that stealth has a significant impact on the aircraft shape.[5]

As a consequence, size and shape of the cockpit are directly dependent on the fuselage design which results from the above considerations. As military fighters usually are required to operate at high speed, which offers many tactical advantages, drag must be minimised. As parasite drag increases exponentially as speed goes up, it is a design goal to keep a fighter's frontal area and fuselage cross-section as small as possible.

On the other hand, fighter pilots refer a lot to visually perceived outside references. As a consequence, an unobstructed outside field of view is essential for a combat aircraft. Large transparent canopies have therefore been adopted in most modern designs. This adds some extra drag, but the advantages of a good all around view by far outweigh the performance penalties. Moreover, the efficiency of modern helmet-mounted displays (HMDs) (see below) directly depend on an unobstructed field of view.

Fighter cockpits are usually much smaller than the pilot workstations of other types of aircraft. Figure 3 shows the front fuselage of the Eurofighter *Typhoon* with its cockpit and the large bubble canopy. The comparison with the body dimensions of the pilot permits an estimation of the size of the cockpit.

With regard to the basic tasks *aviate*, *navigate*, *communicate*, it seems self-evident that the cockpit layout of fighters follows similar patterns as with civil aircraft. As a consequence, the cockpit-T instruments required for a safe conduct of the flight are placed in the best and most central area of the instrument panel. This helps to avoid unnecessary head movements and has a positive effect on pilot workload.

With on-board radar becoming available, an appropriate sensor screen had to be provided. For a long time, a single, monochrome CRT display, suitably located near the pilot's instantaneous field of view, proved to be sufficient for this purpose. In most cases, its size was modest. In some cases, the radar screen was used to alternatively display information from TV-guided munitions or optical sensors. To improve

[5] Cf. for stealth qualities of the F-22 see http://www.globalsecurity.org/military/systems/aircraft/f-22-stealth.htm; Accessed 28 April 2011.

Fig. 3 Single-seat cockpit of the Eurofighter *Typhoon*. The large canopy allows a panoramic outside view. Apart from that, the picture allows a rough estimation of the size of the cockpit

survivability, most fighters also featured a radar warning receiver/display once this technology had reached the necessary maturity. The indicator had to be located in a convenient position so that the pilot could quickly recognise threats and react.

Shortly before the introduction of digital avionics, virtually all fighter cockpits featured the following layout:

- Centrally located, T-shaped arrangement of primary flight instruments,
- Centrally located sensor display (radar screen),
- Other important instruments on the front panel (e.g. fuel, engine indicators),
- Less important controls and indicators on the consoles.

Particularly during low-level flights and air combat, fighter pilots devote a lot of their attention to the outside world, and their actions are often based on visual cues in the environment. Nonetheless they have to keep monitoring flight parameters and system data. With the cockpit layout sketched above (Fig. 2) they had to look down and refer to cockpit instruments frequently, each time re-focussing their eyes from infinity to close range. This dilemma was resolved by the development of the Head-Up Display (HUD) and its early introduction in military aviation. In a HUD, information is projected onto a transparent screen, the so-called combiner, located at eye level in front of the pilot. In the beginning only weapon aiming symbology was displayed. Later, essential information, e.g. attitude, flight parameters, weapons and sensor data, was added. The presentation is focussed at infinity, allowing the pilot to cross check important parameters without forcing him to look into the cockpit and to re-adapt his eyes. While early HUDs provided only a narrow field of view (FOV), state-of-the-art systems offer a more generous size, namely 35° horizontally and 25° vertically, implemented by means of holographic technology. It is important to mention that the HUD optics requires plenty of space. Hence, less room is available for other displays in the centre of the main instrument panel (see Fig. 4).

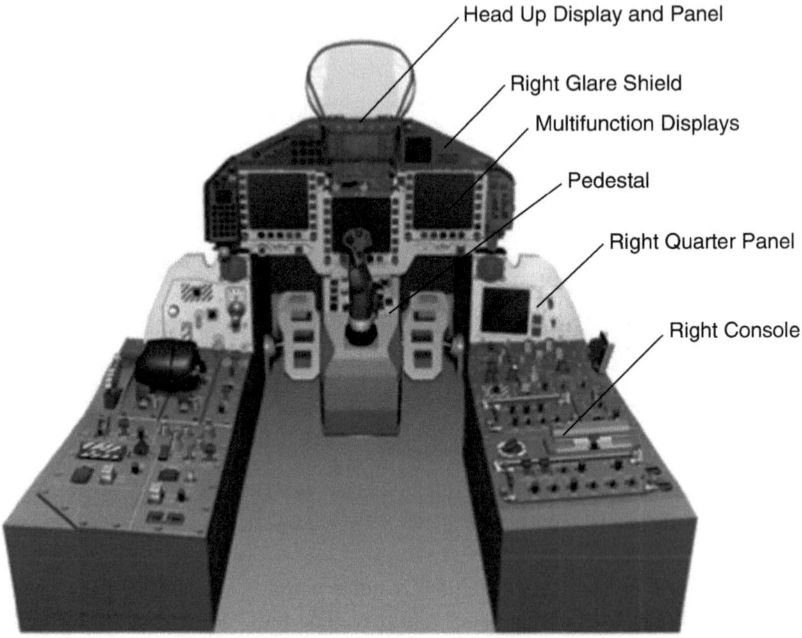

Fig. 4 Eurofighter *Typhoon* Cockpit Layout. The main instrument panel in front of the pilot features the HUD, the head-up panel, three MFDs and the left and right hand glare shields. The left and right quarter panels are installed between the respective glare shields and the consoles (labels are only provided for right-handed components, they apply to the left side in an analogous manner). The primary flight controls pedals, stick and throttle are placed in adequate positions

As the HUD still requires the pilot to look straight ahead while the focus of his interest might be located elsewhere, the next logical step in the evolution was the Helmet Mounted Display. Lightweight construction and miniature electronics enabled the projection of HUD-like symbology onto the pilot's visor or another helmet-mounted optical device. Now the pilot had the information immediately available, regardless of where he was looking. Until today HMDs as well as HUDs provide a monochrome green picture. This is primarily due to the high brightness requirement, which can presently only be achieved by a monochrome CRT. Colour HUDs that generate the image by means of a Liquid Crystal Display (LCD) are currently appearing on the market. More important, aiming and cueing capabilities could be integrated in HMDs. Head trackers measure the viewing direction of the pilot, and it is possible to slew sensors or seeker heads towards a visually acquired target. Likewise, sensor data could be used to direct the pilot's eyes towards an object of interest. While modern fighters use a concurrent mix of HUD and HMD, the Lockheed F-35 *Joint Strike Fighter*, which is currently under development, will be the first combat aircraft relying on the HMD only.

Similar to the HUD, there were specific requirements of fighter aircraft which led to the development of the so-called HOTAS principle (see section below).

It features controls positioned on stick and throttle top allowing the pilot to request specific functions without taking his hands off the primary flight controls.

Another important evolutionary step in cockpit layout became apparent with the introduction of digital computers in the aircraft. They enabled the application of Multifunction Displays (MFDs). Unlike conventional instruments, these electronic display screens offer great versatility and operational flexibility. As a consequence, they allow a much more efficient presentation of complex information. Originally introduced as monochrome displays based on CRT technology, the cockpits of all contemporary combat aircraft feature at least three full colour displays using active matrix LCD technology. Such cockpit architectures are commonly called *glass cockpits* (cf. Jukes 2003).

Figure 4 and 5 show illustrations of the Eurofighter *Typhoon* cockpit. It consists of unique areas: the main instrument panel, the pedestal, the left and right quarter panels and the left and right consoles.

The wide-field-of-view HUD is installed in the upper centre position of the instrument panel. The area below the HUD is part of the pilot's central field of view and hence, from an ergonomic point of view, it is very valuable for information presentation. However, the bulky HUD housing prevents the installation of an MFD, so the area underneath the HUD, i.e. the Head Up Panel is used to control data-link messages and for some essential back-up displays.

The three colour MFDs, each with a 6.25″ by 6.25″ display surface, constitute the largest part of the main instrument panel. The left and right hand glare shield form the frame of the main instrument panel. On the left-hand glare shield, close to the pilot's primary field of view, a panel of system keys is located which provides access to the individual systems. When one of these keys is pushed, the group of moding keys that are positioned next to the system keys is configured to provide access to deeper levels of the selected system. To the left of the left MFD, the Multipurpose Data Entry Facility is installed as part of the left-hand glare shield. It consists of a small alphanumeric display and a keyboard and is also automatically configured to satisfy the specific requirements of the selected subsystem. It enables the pilot to manually enter specific data, e.g. frequencies or waypoint details.

The right-hand glare shield features a set of secondary flight instruments, the so-called Get-U-Home instrumentation. Some of the instruments are covered by a flap, which in turn contains indications related to the data link, transponder and TACAN. The Dedicated Warning Panel (DWP) is placed on the right quarter panel.

Pedals, stick and throttle can be found in ergonomically optimised positions. The latter are equipped with numerous HOTAS control elements.

Controls related to take-off and landing, e.g. the undercarriage lever, are positioned on the left quarter panel.

All other controls for the various on-board systems follow a task-oriented arrangement, i.e. they are arranged in such a way that the functions required by the pilot in a particular situation or during a particular task are placed in the same area in order to minimise the way for the controlling hand. Prior design approaches

Fig. 5 Main instrument panel of the Eurofighter *Typhoon*. The usable panel space is restricted by the aircraft structure, the need for outside vision, and the cut-outs for the pilot's legs. The main instrument panel comprises the left-hand glare shield with the System/Moding Key Panel (**1**) and the Manual Data Entry Facility (**2**), the *left*, *central*, and *right* multifunction head-down display (**3**) and the right-hand glare shield, which includes the Get-U-Home instrumentation (**5**). The Dedicated Warning Panel (**4**) is situated on the right quarter panel

were system-oriented, i.e. all functions belonging to a particular system had been combined on a dedicated system control panel (Fig. 4).

2.3 Cockpit Moding: Information Access and Control Functions

In contrast to civil aviation, military missions follow non-standard, individual routings, which are determined by ad-hoc threat analysis and tactical considerations. Navigation is only a basic means to reach the combat zone where the operational task must be fulfilled in an adverse environment. Ingress into and egress out of the hostile areas are often carried out at very low altitude and high speed to minimise the risk of detection and interception. Virtually all military missions are conducted in formations; four-ship formations, i.e. groups of four aircraft, being the rule. This enhances combat effectiveness by providing mutual support but adds to workload and complexity. To a high degree, the pilots' immediate course of action is determined by sensor information, mainly from the radar. Particularly when air threats are involved, the situation quickly becomes extremely dynamic and unpredictable. Sensors must be managed, weapons employed, information communicated and hostile attacks repelled. Much of modern air combat takes place at supersonic speeds where things happen fast and fuel is scarce.

The pilot has to be provided with exactly the information and the control functions necessary in a particular situation to perform his tasks in the most effective manner. This requirement is fulfilled by a pilot–aircraft interface that features a proper cockpit moding. Referring to the cockpit instrumentation described above, the moding is concerned with **which** information from the installed systems is presented **how** and **when** on the various displays.

Cockpit moding, however, not only deals with information presentation on the displays, it also relates to how the pilot interacts with the aircraft systems by means of specific features such as speech control, facilities for manual data entry, specific key panels or the HOTAS controls (see below).

The complexity of the aircraft systems, the highly dynamic combat environment, and the fact that many modern fighter aircraft are piloted by just one person, require several measures to prepare the available information and control functions for use in the cockpit. Filtering, pre-processing and structuring must be regarded as a means of automation (see below), which serves to keep the pilot workload within manageable limits. Proper cockpit moding is the prerequisite for the pilot to always be aware of the overall situation, to take the right decisions and act accordingly.

The pilot must be provided with all means to build up and maintain situation awareness (SA). According to Endsley (1999) and Wickens (2000) SA is composed of the following components: the geographical situation, the spatial and time-related situation, the environmental conditions, the aircraft systems, and, specifically the military pilot, the tactical situation. Endsley's (1999) model defines three levels of SA, where level 1 relates to the perception, level 2 to understanding and level 3 to the ability of anticipating the behaviour of the system under consideration in a given situation.

SA support in a fighter cockpit is being predominantly related to level 1, i.e. perception of the elements that determine a situation (cf. Endsley 1999). This is achieved by presenting only the relevant information to the visual and aural senses of the pilot. The display of information, which is unimportant in a particular situation, would only affect the pilot's ability to perceive the significant information, resulting in high workload and a drop of SA.

Level 2 SA is supported by the way information is displayed. The presentation has to be immediately comprehensible to the pilot, i.e. the meaning of every symbol must be evident, alone and in relation to other symbols.

There are even features in the Eurofighter *Typhoon* cockpit that aid the pilot in projecting a situation and achieving even SA level 3. The outside vision, but predominantly the HUD, the HMD and the three MFDs are the main windows that provide access to the sort of information that enables the pilot to build up and maintain situational awareness.

2.3.1 Information Hierarchy and Phase of Flight Control

In order to get a complete picture of the situation the pilot needs a task-specific set of formats (the term is explained in the Information Presentation section below)

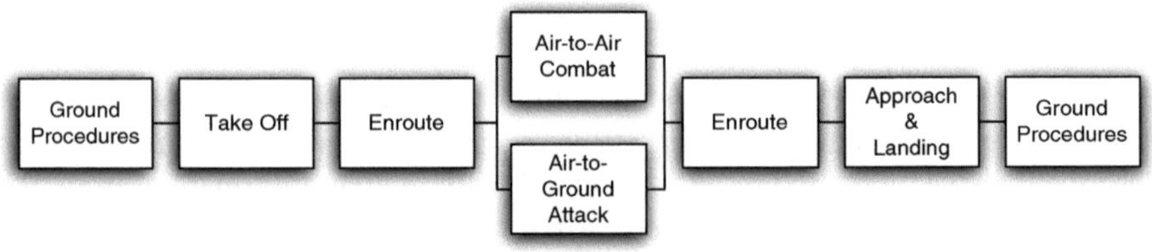

Fig. 6 Mission phases defined as a result of the mission analysis

distributed throughout the entire display suite. To best satisfy the pilot's needs and minimise workload, the formats shown on each of the screens at any time depend on the mission phase. The mission phases, including the required information, have been defined at the beginning of the aircraft development. They are one of the results of a mission analysis which has been performed in order to define the mission with the highest demands, i.e. the design-driving mission, and to establish a data basis for the definition of the cockpit moding. Figure 6 shows the phases identified during mission analysis.

The mission phases represent the highest level of the moding hierarchy. Each of the phases of flight is associated with a particular set of display formats which meets the pilot's information demand for that particular phase. The mission phase is either selected automatically when phase transition conditions are met, or by the pilot by means of a dedicated set of pushbuttons situated on the pedestal. Although the combination of formats usually fits the pilot's needs, he can always select other display formats or any other kind of information he requires for his task. In addition, he has the option to customise the format configuration individually during pre-flight preparation. This can be seen as an example of the adaptable automation philosophy applied to the Eurofighter *Typhoon* cockpit (see below).

Every piece of information available on the cockpit displays has been assigned to one of three levels of a moding hierarchy which is based on importance and frequency of use. Information of the top level is immediately accessible, level 2 information is accessible within a very short time-span, i.e. on one button press, and level 3 information is non-critical, i.e. accessible with a maximum of two button presses. This flat hierarchical structure makes sure that the displays don't become too cluttered and that at the same time, even the remotest information can be accessed fairly quickly.

2.3.2 Information Presentation on the Displays

The information to be presented on the displays is divided into pages called *formats*. To minimise the need to scan other formats whilst performing a task, formats are composed of the information required for that particular task, irrespective of its source.

Each of the three MFDs serves a special purpose. The central MFD is dedicated to the Pilot Awareness (PA) format; this format is displayed in all mission phases.

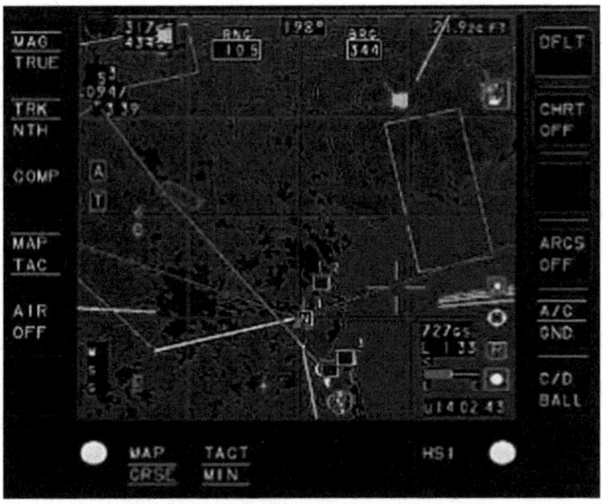

Fig. 7 Eurofighter Pilot Awareness format. Tactical information is displayed on top of a geographically referenced picture. The soft-keys around the display area allow a flexible modification of the display content

On the left MFD in most mission phases, information is presented that complements the PA format by detailed sensor data. The right MFD provides the biggest range of formats. It either shows the situation in a side perspective to supplement the PA format, presents information about the aircraft systems, lists data related to the radios or navigation waypoints or displays warning information in a phase-dependent manner. Swap keys allow interchanging the format packages allocated to each of the MFDs so that any information can be displayed on any display surface.

Figure 7 shows the pilot awareness (PA) format, which exemplifies a typical Eurofighter *Typhoon* display format. It presents navigational data, topographical features and details about the tactical situation on top of a map background. The format is made up of information supplied by different sources such as the Multifunction Information Distribution System (MIDS – see below) and on-board sensors, but it also contains data generated before flight in the course of mission preparation.

Altogether around 20 MFD formats are available in the Eurofighter *Typhoon*, providing in essence system data (fuel, engines, hydraulics), navigation data (maps, waypoints, routings, timings, navigation aids, etc.), weapons data (air-to-air and air-to-surface), sensor data (radar, infrared sensor, etc.), checklists and prompts.

The three MFDs are full-colour displays. Colour is an important feature of modern MFDs since it strongly expands the possibilities for symbol design compared to the older monochrome displays. Due to the number and quality of modern sensors and the networking with other participants present-day military aircraft cockpits are able to present a vast amount of information. Important elements of the displayed information are the so-called tracks, dynamic representations of objects in the tactical scenario. The data available about a track is displayed as track attributes. One of the crucial attributes is the allegiance of a track. The pilot must be able to recognise at a glance whether a track is hostile, friendly or unknown. This attribute is communicated in the *Typhoon* by colour and shape, i.e. double coded to support the perception under critical lighting conditions. Other attributes are coded, e.g. by the outline of a symbol or its size. On top of that there is other track related data, continuously displayed or selectable, which is arranged around the relevant track symbol.

One criterion for the distribution of information to the various display surfaces is the pilot's need to have the relevant information available either head up while directly observing the environment or head down, e.g. during extensive planning or diagnostic activities. A big portion of fighter operations is associated with the necessity for the pilot to visually observe the air space. In order to provide him with the most basic information even in these situations, HUDs and HMDs have been introduced. They can present flight reference data as well as specific information from the on-board sensors and weapon aiming information. With regard to head-down operation it should also be mentioned that essential flight data is replicated on the tactical displays to limit the need for extensive crosschecks whilst staying head-down and hence reduce the information access effort.

2.3.3 Manual Interaction and Information/System Control

Manual control of the display content is being performed by a set of 17 pushbuttons arranged on three sides of each of the MFD screens (see Fig. 5). They facilitate the selection of other formats and the interaction of the pilot with the currently displayed format. The interaction options depend on the type of format. The configuration possibilities of the PA format, for example, include several declutter modes, the map orientation, i.e. north up/track up presentation, map on/off, etc. These so-called *programmable soft-keys* are a peculiarity of the Eurofighter *Typhoon* cockpit. They embody alphanumeric displays, which provide a dynamic indication of their functions and their current settings. Figure 8 presents a detail of a Eurofighter *Typhoon* MFD showing the bezel with two of these soft-keys. The right one, for instance, allows selecting the HSI (Horizontal Situation Indicator) format. The advantage of these soft-keys is that no space has to be reserved on the screen to indicate the button functions and its menu options, unlike the plain buttons found in other aircraft. Hence the display surface can be used to its full capacity for the indication of other essential information, and the pilot's perception of the

Fig. 8 Detail of a Eurofighter MFD showing two soft-keys, i.e. programmable pushbuttons that feature an alphanumeric display to show their function and status

information is not distracted by button captions. This can be understood as situation awareness support on level 1.

Apart from that, the soft-key technology conforms best to the spatial compatibility requirement, which results in a better reaction time (Wickens and Hollands 2000). And finally the indication of the system states make sure that the pilot is kept in the loop, which in turn improves his situation awareness on levels 1 and 2.

Soft-keys are not the sole possibility to interact with the display; other options are the xy-controller and speech control. The xy-controller, situated on the throttle grip, thus being part of the HOTAS controls, allows moving a cursor over all three MFDs and the HUD to control the presentation. It is used, for example, to present more detailed information on a selected track, such as speed, height and type, but also to increase/decrease the range of the format.

2.3.4 HOTAS

HOTAS is an acronym for Hands On Throttle And Stick. It stands for a control principle applied in most modern fighter aircraft. When they are engaged in air combat, the pilots are regularly exposed to high centrifugal accelerations, G-forces that push the pilot into the seat. These extreme dynamic manoeuvres are flown manually, which means it is impossible for the pilot to leave his hands from the primary controls stick and throttle. In parallel, even in such situations he has to prepare and control the weapon release. As a consequence, the grips of stick and throttle are designed in such a way that essential functions can be directly controlled by means of switches mounted on these grips, having the advantage that the pilot need not move his arms under such high-G conditions in an attempt to operate a control e.g. on the main instrument panel. This arrangement also enables the pilot to keep his eyes on the outside world whilst interacting with sensors and weapons, which is very efficient in connection with the use of the HMD. HOTAS controls are given specific shapes and textures to allow the pilot a haptic distinction thus enabling him to operate them without visual reference.

The Eurofighter *Typhoon* features more than 20 different controls on the throttle and stick tops. HOTAS concepts usually avoid the use of switches for multiple functions in order not to confuse the pilot. However, swing role missions, which imply a rapid change between air and surface engagements, employing different kinds of sensor modes and weapons, have caused a double assignment of a very limited number of HOTAS controls. Bearing this in mind, it is not surprising that proper handling of complex HOTAS switches requires extensive training and frequent practice.

2.3.5 Speech Control

An additional means of control in the *Typhoon* cockpit is speech control. It allows the pilot to interact with the aircraft systems via spoken commands as an alternative to manual control. Similar to HOTAS controls, speech control is an isotropic command method, i.e. it is independent from the pilot's viewing direction and therefore ideal for head up operations. Since no HOTAS control is replicated by a spoken command, speech control can be considered an extension of the HOTAS concept. It is used to control a variety of cockpit functions, e.g. on the Manual Data Entry Facility (MDEF – see below) or the display soft-keys. A limiting factor for its application is the achievable recognition probability. Since speech control is not perfectly reliable, the commanding of safety-relevant functions by speech is excluded.

2.3.6 Left-Hand Glare Shield

Many avionic subsystems are controlled by means of a facility located on the Left-Hand Glare Shield. Close to the pilot's central field of view there is a key panel which provides access to the individual systems via fixed function keys, the system keys, and an array of multi-function keys, which serves to control the modes of each subsystem, the moding keys (see Fig. 5 (1)). The moding keys have the same design as the soft-keys on the MFDs. When a system key is pushed, the moding keys configure to allow control of the selected subsystem.

Another element of the Left-Hand Glare Shield is the Manual Data Entry Facility (MDEF) situated to the left of the left MFD (see Fig. 5 (2)). It consists of an alphanumeric display featuring four read-out lines and a keyboard that appears much like the one of a mobile phone. The read-out lines and the keyboard automatically configure to satisfy the specific requirements of the system selected for data entry. Selection of a radio, for example, automatically configures the MDEF for frequency entry.

2.3.7 Reversions

Several mechanisms make sure that the pilot is still provided with the best information available should one or more displays fail. Directing the formats of the failed display to a working screen can compensate for failures of one or more MFDs. The Display Swap Keys allow interchanging the display content of two MFDs thus enabling the pilot to access any information in case of display failure. Should the HUD fail, the relevant information can be shown on an MFD, and also the Get-U-Home instruments on the Right-Hand Glare Shield can be used for flight guidance. If the display suite fails completely, the Get-U-Home instruments will remain available.

2.4 *Automation*

2.4.1 General Remarks

The number and variety of pilot tasks in handling the complex on-board systems in a highly demanding tactical environment lead to a degree of automation which makes sure that the workload of the pilot is kept on an acceptable level. Since "automation can be thought of as the process of allocating activities to a machine or system to perform" (Parson 1985), it is always associated with taking the human operator out of the loop, which makes it increasingly difficult for the operator to refresh his knowledge about changing system states. The so-called *out-of-the-loop unfamiliarity* refers to the degraded ability to discover automation failures and to resume manual control (Endsley and Kiris 1995). This can cause severe problems, particularly with regard to the interaction between operator and the automated system (cf. Lee 2006). It was found that, as a result of the automation, the operator is often confronted with an unpredictable and difficult-to-understand system behaviour, which in turn leads to automation surprises and so-called *mode confusion*.

In order to avoid the emergence of out-of-the-loop unfamiliarity it is important that the pilot always holds a correct mental model of the aircraft and its condition (cf. Goodrich and Boer 2003). Therefore, in some cases it is essential that the pilot is able to overrule the results of automation, following an adaptable automation approach as described by Lee and Harris et al. (Lee 2006; Harris et al. 1995). The pilot may engage or disengage the automation as needed, which gives him additional degrees of freedom to accommodate unanticipated events (Hoc 2000). However, the decision and selection process is an additional task and may increase the pilot's workload, mainly in demanding situations.

Automation can support the pilot in basically any part of human information processing, i.e. perception, cognition and action. Proper preparation, processing and filtering of information assists perception and can be considered an automation feature just as decision support measures that foster cognition or automatic performance of defined tasks, which relieves manual control.

In the Eurofighter *Typhoon* cockpit there are plenty of automated functions that alleviate the task of the pilot. A prominent example, which is also common with civil aviation, is the autopilot/autothrottle. It causes the aircraft to acquire and hold a certain parameter (e.g. altitude) or a combination of parameters (e.g. altitude and heading) or to follow a predefined flight path automatically when selected by the pilot. The outcome of this is a release from the manual steering task, which leaves the pilot more capacity for mission management.

2.4.2 MIDS

Today most military entities participating in a co-ordinated operation use digital information network services to exchange data in nearly real time. The system employed in the NATO is called MIDS (Multifunctional Information Distribution System) with its tactical data network service LINK16.[6] It provides a variety of information of different type and character, partially directly related to the formats displaying the tactical situation, e.g. the positions of tracks. This data is updated automatically on the format whenever it is received. Other information has the form of alphanumeric messages, e.g. commands or requests from the co-ordinating agency. They are displayed at the time of arrival on the most appropriate display or on the HUD Control Panel, where the pilot can also select from a list of pre-defined responses. Information organised as data lists, e.g. waypoints, which serve to update the originally loaded data, are directed to the respective system at the time of arrival. The reception of new data is indicated to the pilot. The updated lists can be displayed upon selection on one of the MFDs.

2.4.3 Sensor Fusion

The Eurofighter *Typhoon* features several sensors which provide track identification and, whenever available, positional data to the avionics system. Unfortunately, the data quality, timeliness and accuracy of these sensors – LINK16 is treated as one of them – differ considerably in various dimensions, i.e. azimuth, elevation, and range and also with regard to the identification data. If the track data of the sensors were displayed individually, this would lead to a very cluttered display, since identical tracks would be indicated by two, three or even four symbols, one representing each sensor. Since such a situation is not acceptable for the pilot, the data from the various sensors are correlated and fused, so that each track is represented by one consistent symbol on the display format. The underlying sensor fusion algorithms are very complex but the result is a much clearer tactical picture. From a situation awareness perspective, this automatic feature must be deemed to be critical, as, in order to build confidence in the displayed information and to be able to check the indication for plausibility, it is essential for the pilot to know which sensor data a particular track is formed of. This is the only way for him to establish a mental model about the functioning of that feature. Experience has shown, that on the one hand sensor fusion supports the perception of information, i.e. SA on level 1, but on the other hand it is very challenging to provide an implementation which holds the pilot in the mental loop and supports SA on level 2 (see Endsley 1999). Currently, the symbols indicating fused data show the data sources by corresponding symbol coding and an indication of the identification confidence to allow the pilot a judgment about the validity of the data. Apart from that, the algorithms can be adapted over a wide range to cover peculiarities of the sensors or to account for different rules of engagement.

[6] Cf. http://de.wikipedia.org/wiki/Multifunctional_Information_Distribution_System; Accessed 29 May 2011.

2.4.4 Warning System

Another feature, also known in civil aviation, is the dark cockpit philosophy. It means that during normal operation when all systems are properly working, the pilot does not receive any information about the state of his aircraft, so he is not required to monitor the state of the systems or equipment. This principle saves display space, which can be used for more important information hence it reduces workload. Only in case of a malfunction the pilot is informed by the warning system that something is out of order; problems can thus be easily identified. The warning system alerts the pilot, provides multimodal information as to what that warning situation is, gives advice about necessary actions and informs about the potential consequences of an unserviceable system.

When a failure occurs, it is the first function of the warning system to draw the pilot's attention to the failure situation. Attention getters start flashing and a so-called attenson (attention getting sound) sounds. The attention getters are pushbuttons with an integrated red light placed in the pilot's primary field of view, to the left and right of the HUD (see Fig. 5). Their sole purpose is to alert the pilot. After a few seconds the attention is replaced by a voice warning, e.g. "left fuel pressure", which is used to inform the pilot as to the exact nature of the warning. At the same time the relevant caption ("L FUEL P" for left fuel pressure) on the Dedicated Warning Panel (DWP) (see Fig. 5) flashes. All flashing and the voice warning continue until the pilot acknowledges that he has noticed the warning by pushing one of the attention getters. The DWP caption becomes steady and the pilot can take diagnostic and corrective action.

Failure captions on the DWP are located in "handed" positions, i.e. failures of systems that are at the left aircraft side are indicated in the left column of the DWP captions. Captions are either red or amber to indicate the criticality of the problem. With every additional failure the warning sequence starts again but the captions on the DWP are re-arranged according to the priority of the failure. Further advice is available on the MFDs, where the pilot can retrieve information about relevant aircrew procedures and the potential consequences of a system malfunction on aircraft performance.

The alert and inform functions of the warning system support situation awareness on level 1. Their mechanisation is realised via the attention getters, the attenson, the voice warning and the DWP. Level 2 and 3 SA (Endsley 1999) are enhanced by the display of the procedures and consequences pages.

The warning system, however, is not only used to indicate system and equipment failures. It is also designed to alert and inform the pilot about hazardous situations that may occur during aircraft operation and that require immediate action. Such situations may, for example, be undue ground proximity or the approach of a hostile missile announced by the Defensive Aids Subsystem. In those cases the warning system generates appropriate voice warnings, e.g. "PULL UP" or "BREAK RIGHT MISSILE" to prompt the pilot to perform an evasive manoeuvre.

2.5 The Cockpit's Future

The presence of military aviation is determined by the discussion whether UAVs will be able to take the role of manned fighter aircraft in the next couple of years. There are even voices in the military aeronautical community that challenge the future of manned aviation beyond the latest generation of fighter aircraft in general. Given the situation that some mission types, predominantly reconnaissance and surface attack missions, are already operated by UAVs it is certain that the rate of manned operation in those missions will decrease. For air combat, however, characterised by highly dynamic manoeuvring, manned aircraft are still the first choice. This is one reason why military aviation may also have a manned component in the future. The other reason is that history shows that military aircraft have an in-service life of 50 years and more, which usually involves multiple modernisations to keep their combat value up-to-date and to compensate for obsolescence issues. Most of the necessary modifications will also affect the cockpit, as basically any aircraft system is represented in the cockpit and it is necessary to make sure that the pilot's workload is kept on an acceptable level and his situation awareness is not degraded despite an increase of functionality.

In recent years some new aircraft developments have originated, the latest being the Lockheed F-35, *Joint Strike Fighter*. Its cockpit revokes the principles that have determined the cockpit layout of fighter aircraft for a long time. It does no longer feature a HUD and its main instrument panel is dominated by a single display surface of $8'' \times 20''$. Head-up operations are carried out using a HMD only.

The trend towards an increased display space can be observed since the first multifunction displays have found their way into the cockpits of fighter aircraft and it still continues. In present generation aircraft, AMLCDs (Active Matrix Liquid Crystal Displays) with a size of $6'' \times 8''$ are state of the art. The F-35, however, with its $8'' \times 20''$ display area opens up the way towards an implementation of the ideas of Hollister et al., who propagated the *Big Picture* in the fighter cockpit (Hollister et al. 1986) in the 1980s already, when the technology for an implementation was not yet available. Lockheed achieves this display area with two $8'' \times 10''$ AMLCDs, which are installed side-by-side nearly seamlessly (Ayton 2011).

Within EADS quite a different display concept has been investigated, which is based on back projection (Becker et al. 2008). This technology allows building modular and scalable displays which can be adapted in a very flexible manner to the size and shape of the main instrument panel of existing aircraft and are therefore particularly suitable for cockpit upgrades. A holographic screen improves the energy efficiency and the optical quality of the display at the same time.

Utilising the full potential of a display screen which offers much more space than current cockpit concepts, however, requires a completely new approach with regard to pilot interaction. Kellerer et al. (2008) developed a display concept which presents an integrated pilot awareness format all over the screen, allowing the pilot to select specific perspectives on the tactical presentation according to the current situation. Small windows can be superimposed, permitting direct system information access upon pilot demand (see Fig. 9).

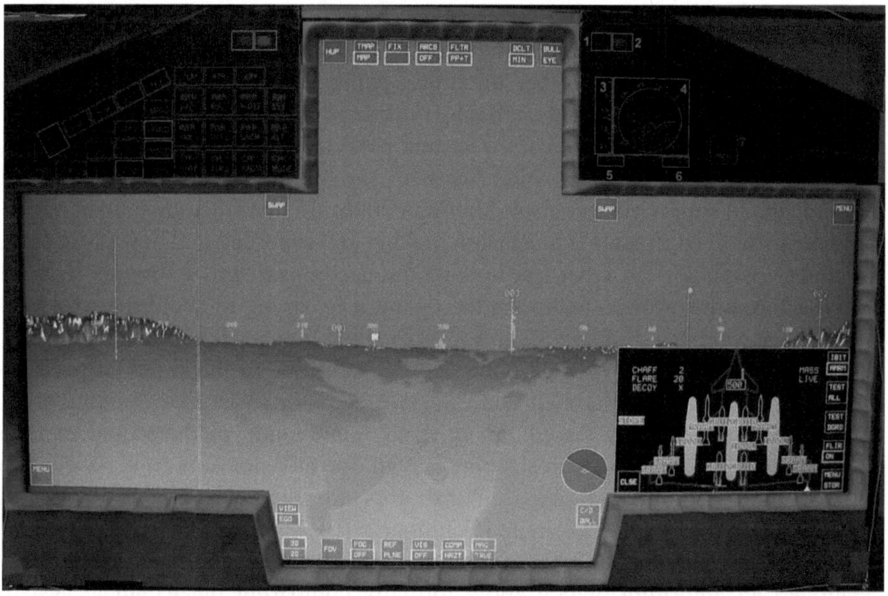

Fig. 9 Display concept according to Kellerer et al. (2008) showing a full screen integrated pilot awareness presentation overlaid by a small special purpose window, which can be selected and deselected if necessary

One implication of such big displays is the lack of space for conventional hardware controls such as pushbuttons or rotary controls on the main instrument panel. The F-35 designers overcome this issue with a touchscreen based on infrared technology and with speech control as isotropic input medium (Ayton 2011). Touch screens not only allow to show interactive areas on the screen mimicking conventional pushbuttons, but also to interact directly with other graphical elements on the screen. Kellerer (2010) and Eichinger (2011) found a significant performance advantage of touchscreen interaction compared to non-direct interaction via cursor control devices, such as a track ball. Their theoretical and experimental results give convincing evidence that this will also be feasible in the harsh environment of a fighter cockpit. In addition to the touchscreen, their control concept foresees a completely redundant xy-controller on the throttle top, being an element of the HOTAS controls, completed by speech control as a third input medium.

Considering the ever-increasing functionality in the fighter cockpit, automation, capable of sustainably unburdening the pilot, will be another area of improvement in the future. One of the approaches investigated in aeronautical research in recent years is adaptive automation. Adaptive automation can be understood as an advancement of adaptable automation and, according to Lee (2006), implies an automatic adjustment of the level of automation based on the operator's state. The basic idea of adaptive automation is avoiding both an overload and an underload of the operator. The challenge consists in the exact detection of the operator's state. Most physiological parameters react slower than required, and the measured values are not always consistent enough to be qualified for a real application, particularly in the environment of a fighter aircraft. Irrespective of the complexity of the measurement, there are two other important points, which have to be taken into consideration: the measurement itself may impose additional workload (Kaber et al. 2001) and the dynamic changes of the system potentially induce automation errors.

As mentioned above, automation tends to take the human out of the loop and often results in supervisory control (Sheridan 2006). In that role, the human operator is only involved in high-end decision making and supervising the work process, which, at least in normal, error-free operation is accompanied by a strongly reduced workload. Another way to unburden the operator by trying to avoid the workload dilemma, is *cognitive automation*, i.e. providing human-like problem-solving, decision-making and knowledge-processing capabilities to machines in order to obtain a goal-directed behaviour and effective operator assistance (Schulte et al. 2009). Such systems are based on artificial intelligence methods and can be understood as machine assistants supporting an operator not only in his executive but also in his cognitive tasks. Demonstrators are already available in several research laboratories (see Onken and Schulte (2010), Sect. 5.3.1).

Degani and Heymann (2002) follow a quite different approach. They have developed an algorithm, which allow formally comparing the technical system and the mental model of the operator and hence discover ambiguities and errors in the automated system. Specifically, they model the technical system as well as the operator's mental model of the system using the state machine formalism. The latter is derived from training and HMI information material. The formal comparison reveals mismatches between the two models, which indicate either errors in the automation or deficiencies in the mental model of the operator. They can be eliminated by changing the automation, the training, or the interface (Lee 2006).

3 UAVs: Characteristics of the Operator Interface

The fundamental difference between UAVs and manned aviation is that the person who guides the aircraft, i.e. the platform, is not sitting in a cockpit but in a control station anywhere on the ground. So in a literal sense it is not a truly unmanned system, the term "uninhabited" or "unmanned" is exclusively related to the flying component. The flying platform alone, however, is not sufficient for UAV operation. Therefore the notion UAS (Unmanned Aircraft System) is often used. UAS is also the official FAA term for an unmanned aerial vehicle in order to reflect the fact that these complex systems include ground stations and other elements besides the actual aircraft. The inclusion of the term aircraft emphasizes that, regardless of the location of the pilot and flight crew, the operations must comply with the same regulations and procedures, as do those aircraft with the pilot and flight crew on board. A UAS consists of the UAV, the ground station, the data link, which is connecting the UAV with the ground station, and finally one or more human operator(s). So, as one or more operators are always in the loop in some way, human factor aspects also have to be considered in the UAV domain in order to assure successful operation.

Unmanned systems are generally seen to take over the tasks and missions that are too *dull*, for instance extended mission times of sometimes more than 30 h, *dirty*, e.g. exposure to contaminations, or *dangerous*, e.g. human exposure to hostile threat for manned aircraft. Due to the lower political and human cost in the case of a lost mission, they are an appreciated alternative to manned flight. Humans are often the limiting factor in performing certain airborne roles. An unmanned aircraft is not limited by human performance or physiological characteristics. Therefore, extreme persistence and manoeuvrability are intrinsic benefits that can be achieved with UASs (US Air Force 2009). Moreover, if a UAV is shot down there is no pilot who can be taken hostage.

The absence of a human crew results in a number of advantages for the construction of the flying platform: less weight since no pilot including his personal equipment, cockpit, cockpit avionics, life support and crew escape systems have to be integrated, and a technical construction of the same flight performance which is much smaller, simpler, cheaper and less fuel-consuming as no external vision requirements have to be considered. A reduction of development, training, support and maintenance costs is also mentioned frequently when the advantages of UAVs over manned aircraft are discussed (Carlson 2001). The most important disadvantage is seen to be the particularly high requirements concerning a reliable communication link between UAV and ground station.

At present there are a great number of producers who are developing UASs. Within the scope of a study La Franchi (2007) listed 450 different unmanned programmes. In terms of cost, existing UASs range from a few thousand dollars to tens of millions of dollars; in terms of capability, they range from Micro Air Vehicles (MAV), weighing less than 500 g, to aircraft weighing over 20 t (Office of Secretary of Defense 2005). Another figure suitable to appreciate the relevance of UASs in current military operations is the following: as of October 2006, coalition UASs, exclusive of hand-launched systems, have flown almost 400,000 total flight hours in support of *Operation Enduring Freedom* and *Operation Iraqi Freedom* (Office of Secretary of Defense 2007).

Although multi-role airframe platforms (cf. mission diversity in the introductory section) are becoming more prevalent, most military UAVs fall into one of the two categories: reconnaissance/surveillance (providing battlefield intelligence) and combat (providing attack capability for high-risk missions). In recent years, their roles have expanded, but ISR[7] are still the predominant types of UAV missions, particularly when operated by European forces.[8]

[7] Intelligence, Surveillance, and Reconnaissance: "An activity that synchronizes and integrates the planning and operation of sensors, assets, and processing, exploitation, and dissemination systems in direct support of current and future operations. This is an integrated intelligence and operations function" (Joint Chiefs of Staff 2010).

[8] Currently, only the United States and Israel are known to have employed weapon carrying platforms (http://defense-update.com/features/du-1-07/feature_armedUAVs.htm; Accessed 28 July 2011).

The sensors used in such missions are primarily cameras operating in the visual and in the infrared spectrum. Synthetic Aperture Radar (SAR) sensors are also common. The sensor pictures/videos should preferably be transmitted in real time.

UAVs can be classified according to the manner they are guided. They are either remotely controlled manually like model airplanes, where the *pilot* directly affects the aerodynamic control surfaces via radio signals, or their guidance is highly automated, i.e. the *pilot* has no direct control, he can only modify the inputs for the autopilot. There are also UASs whose kind of guidance depends on the phase of flight. On some UASs take-off and landing are performed manually by an *external* pilot (Manning et al. 2004). In any case operation of the UAV depends on a working data link, primarily to downlink status information such as position, speed or altitude from the UAV to the ground station, but also to be able to uplink guidance commands to the UAV (Buro and Neujahr 2007).

3.1 Data Link: The Critical Element

The data link has certain characteristics which have an impact on the design of the UASs and in particular on the design of the human–machine interface. This applies to the guidance of the flying platform but also to the control of an optical sensor on board the UAV:

- The data link has limited transmission capacity: all the information the pilot of a manned aircraft is able to perceive, just because he is physically residing on the platform (e.g. visually from the view out of the cockpit, acoustically or even olfactorily e.g. the smell of fire), would in case of a UAV have to be picked up by a technical sensor and be sent to the ground station. Due to the limited capacity of the data link, such an effort is usually not taken. Normally, the UAV commander is only supplied with information about the status of the aircraft and its systems. More advanced systems are able to display to the UAV commander the picture of an on-board camera which corresponds to the view out of the cockpit, thus improving his situational awareness.
- It is not free of latencies: latency refers to the time delay between the UAV operator making a control input and the feedback received by the operator from the system to confirm that the control input has had an effect. These latencies can account for a second or more, particularly when satellites are used to relay signals (Boucek et al. 2007). Direct feedback, essential for proper manual control is thus not assured. For systems where the distance between the ground station and the UAV causes a considerable time delay, direct manual remote control cannot be recommended because the associated latencies require a high degree of the operator's predictive capability and make control highly error-prone.
- It has a limited range: a terrestrial data link requires a line-of-sight (LOS) connection between the antennae on the ground and on board the platform. Hilly

or mountainous terrain may considerably limit the range. Beyond-line-of-sight (BLOS) data links, which utilise satellites, can find a remedy; however at the expense of longer transmit times.

- It is not interruption-proof: interruptions of the data link can be caused e.g. by topographical obstacles (mountains) in the direct line-of-sight, the data link can be jammed by third parties, but also atmospheric disturbances may cause irregularities of the data transmission.

These characteristics have an impact on the design of the operator interfaces, be it for the commander or the sensor operator (the tasks of the operators are explained in the section The Roles of the Operators and the Human–Machine Interface below). First of all, they affect the operators in maintaining optimum situational awareness, since the supply of complete and up-to-date information cannot be guaranteed. Secondly, in case the link is lost they are not able to intervene and send instructions to the UAV (see Manning et al. 2004).

Therefore, arrangements must be made to avoid that the data link properties affect the safe operation of the UAV. In modern systems, the latency problem is tackled by a design that relies on automatic guidance. The UAV is controlled by means of high-level commands (e.g. *Take Off* or *Return to Base*), which are complex instructions for the UAV's autopilot. Interruptions or even total losses of the data link, however, require special design methods. Since predictability of the flying platform behaviour is of paramount importance in order to warn or re-route other aircraft in that area, the flight control computer on board the UAV, in case of a data link loss, holds a programme, which is automatically activated when the link is interrupted for a certain time span. One method to cope with data link interruptions is to command the UAV to a higher altitude in order to try to re-establish the line of sight and thus allow the data link to resume operation. In case a reconnection is not possible, an automatic return to the airbase is another option. Whenever an unguided UAV, uncontrolled due to data link loss, is flying in the airspace, there must be an authority that makes sure that no other aircraft collides with the UAV. Air traffic control, both civil and military, takes over that role.

3.2 Level of Autonomy

Since the UAV has to show a predictable behaviour once there is a loss of the command data link, it has to be able to operate autonomously, at least in certain situations. Autonomy is commonly defined as the ability to make decisions without human intervention. That complies with automation level 10, the highest on the levels-of-automation scale created by Sheridan and Verplank (1978), which says: "the computer decides everything and acts autonomously, ignoring the human". To that end, the aim of autonomy is to teach machines to be *smart* and act more like humans. To some extent, the ultimate goal in the development of autonomy technology is to replace the human operator or at least minimise the degree of his involvement. Study Group 75 of the NIAG (NATO Industrial Advisory Group) defines four levels of autonomy (NIAG 2004):

Level 1	Remotely controlled system	System reactions and behaviour depend on operator input
Level 2	Automated system	Reactions and behaviour depend on fixed built-in functionality (pre-programmed)
Level 3	Autonomous non-learning system	Behaviour depends upon fixed built-in functionality or upon a fixed set of rules that dictate system behaviour (goal directed reaction and behaviour)
Level 4	Autonomous self-learning system	Behaviour depends upon a set of rules that can be modified for continuously improving goal directed reactions and behaviours within an overarching set of inviolate rules/behaviours

Regardless of whether level 3 or level 4 autonomy is aimed at, there is still a long way to go before we are technically able to achieve these levels. Largely or completely autonomous systems, level 7 and higher on the Sheridan & Verplank scale (1978), even if they are technically feasible, are problematic from a legal point of view, since there must be a person, the pilot-in-command of a UAV, who can be held accountable should an accident happen. Machines are not able to assume responsibility. This aspect has also far-reaching consequences for the design of the ground station and the operator interface. The acceptance of responsibility implies freedom of action and freedom of choice. Situations in which an operator has to act under duress usually release him from responsibility, unless he has brought about this situation deliberately or negligently (Kornwachs 1999). Therefore, the responsible UAV operator must not be brought in a situation where he is unable to act freely, which may be the case when the ground station does not offer the necessary possibilities for information and for action.

Today, the scope of guidance options for UAVs extends from manual remote control to the widely automatic following of a programmed flight plan. Many of the UAVs in operation today feature a phase of flight-dependent level of automation: take-off and landing are manually controlled while the rest of the flight is automatic. Unfortunately take-offs and landings that are manually controlled are very accident-prone. More than 60% of the accidents occurring in these situations can be attributed to human error (Williams 2006). In order to resolve this issue the UAVs are given a higher degree of automation, which enables them to perform take-offs and landings automatically, thus achieving a much higher safety level. This may be called autonomy, but it is autonomy with regard to the flight control function only. Flight control no longer depends on pilot input, at least as long as the operation is free from defects. A change of the programmed behaviour of such systems is only possible via complex "high-level" commands – this can be compared with the pilot of a conventional aircraft who is only changing the modes of an autopilot, which is never disengaged.

An autonomous flight control system as described, however, is not the only potential area of far-reaching automation. Other concepts, e.g. for automatic sensor control or a combination of sensor control and flight behaviour are already existing.

3.3 *The Roles of the Operators and the Human–Machine Interface*

Considering the tasks of an aircraft pilot (see *Introduction*) and comparing them with the duties of a UAV commander of a modern, highly automated UAV which features automatic guidance, it can be noticed that the **aviate** task has been taken away from the operator. Flying skills do not play any role in UAV operation today. The other tasks, i.e. navigate, communicate and the management of the resources and the mission still apply to UASs. They are usually split up into two categories of operators who fulfil two basic functions: **UAV guidance**, with the UAV commander being the responsible person, and **payload control**, predominantly all sorts of sensors, which is carried out by the payload operator. Both functions are characterised by the physical separation from the flying platform and the interconnection of the operators with the platform through a data link as described above.

UAV guidance in modern systems is mainly associated with monitoring the progress of the mission and with only little interaction, e.g. sending the *Take-Off* command. Only in case of unpredicted events there is a need for the UAV commander to take action. Technical problems or the necessity to modify the pre-planned route due to military threats, weather phenomena or other air traffic are typical examples of such events. One critical question that arises in that context regards the type of information that should be presented to the UAV commander and the way it should be displayed. It is adequate to show the primary flight information (basically attitude, speed, altitude and heading) of conventional aircraft (the Basic-T) or, considering that the commander is not able to directly steer the UAV, is it sufficient to design an HMI which much better fits in with the requirements of a typical supervisory task, such as that of an air traffic controller?

Most UAV commanders in the armed forces and at the manufacturers' are fully-fledged current or former pilots. Their education and experience are essential because the UAVs operate in the same airspace as manned aircraft and UAV operation is subject to the same rules and regulations as conventional flight. It requires knowledge of the air structure, the applicable procedures and a pronounced spatial understanding. In order to make appropriate decisions, the UAV commander has to put himself into a position as if he would be on board the platform (Krah 2011).

Discussions with UAV commanders indicate that the feedback they are experiencing in a real cockpit is missing in a UAV ground station. Particularly, the lack of acoustic and haptic sensations, but also the absence of an external view makes it difficult to imagine being in a cockpit situation. They feel more like

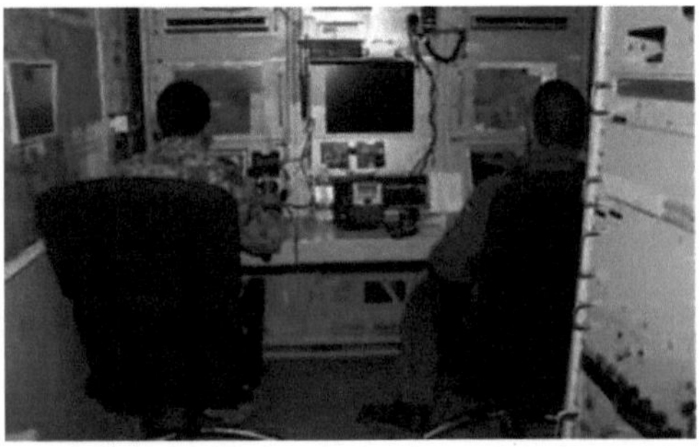

Fig. 10 Heron ground control station (Source: Krah 2011)

observers than acting persons and they criticise missing immersion, which affects their ability to generate appropriate situational awareness.

At the Farnborough Airshow 2008 a concept was presented, called the *Common Ground Control System*,[9] with a view to overcome at least the visual feedback issue.

This idea is applicable to basically any type of UAV, whether remotely piloted or automatically guided. By presenting an external view to the operator like the one he would experience when looking out of the cockpit window, he can be provided with a much clearer picture of the situation than ground stations currently in operation, such as e.g. the Heron Ground Control Station (see Fig. 10) usually do. The benefit for the operator of an external view, provided by on-board cameras, is not only to get a broader picture of the UAV environment, but also to be able to see the impact of e.g. turbulence on the flight.[10] To a certain degree this may compensate for the lack of haptic feedback. However, such a design implies the availability of a powerful down-link.

The ground control stations of most known UASs provide the UAV commander with the indications and controls typically available in conventional aircraft: primary flight information, a map display, aircraft system status indications, warnings, controls for radio communication, etc. and, for those systems that are highly automated, few controls that allow interaction with the UAV. The UAV commander, if he has a pilot education, is already familiar with this cockpit-like human machine interface, thus it is commonly considered appropriate for UAV guidance. Whether it is optimal to fulfil the task, however, has not been clarified so far.

The **payload control**, i.e. the handling of the (optical) sensor is the task of the payload operator, who also has to communicate with the recipients and users of the picture and video sequences, in order to position and adjust the cameras according to the demands. Payload operators, at least in Germany are recruited from the group of experienced picture analysts (Krah 2011).

---

[9] See figure (http://www.raytheon.com/newsroom/technology/stellent/groups/public/documents/image/cms04_022462.jpg; Accessed 30 August 2011).

[10] Turbulence is much quicker perceived by observing a camera picture than just a small attitude indicator.

As a basic principle, the workstation of a sensor operator comprises two displays, one to show the sensor picture and another one for geographical reference, the map display. In designing a payload control station, which facilitates efficient task fulfilment, a number of aspects, have to be taken into account:

- The sensor operator is not only separated from the platform but also from the sensor, so the designer and finally the user have to deal with different reference systems,
- Being not part of the flying platform makes it more difficult for the operator to maintain situational awareness and
- The latencies between command and visible system reaction affect proper manual sensor adjustment.

Keeping track of the mission is sometimes critical. If the sensor is used to search for objects, i.e. flight direction and sensor line-of-sight are not identical, the initial field of view is relatively large, covering a big geographical area and providing a good overview. After an object has been detected, the *zoom-in* function in conjunction with geo-stabilisation will be used to investigate the details. The smallest field of view represents a keyhole view, which is associated with a high probability for the operator to lose track. Also while searching with the sensor it is very demanding to keep track during turns, i.e. to correlate the camera picture with the flight path. Buro and Neujahr (2007) have compiled a number of requirements for the design of a sensor control HMI; the most significant ones are:

- The operator needs an indication that predicts the effects of his control commands
- Sensor adjustment shall be possible through direct cursor positioning in the sensor video
- There shall be an indication of the next smaller display window (zoom-in)
- The operator shall be provided with a synthetic view onto the area around the sensor window for better situational awareness (see Fig. 11).

Another feature able to support the situational awareness of the payload operator is a sensor footprint on the map display.

3.4 Sense and Avoid

Legal regulations currently restrict the operation of UASs to a specially designated airspace. Access to the public airspace, which would be a prerequisite for large-scale military and civil operations, is refused due to the fact that UAVs do not obtain the safety standard of manned aircraft. The critical aspect here is their missing capability to avoid collision with other air traffic in the immediate vicinity. In manned aviation, the pilot is considered to be the last instance for collision avoidance since he is able to see other objects and react accordingly. This is referred to as the *see-and-avoid* principle according to the FAA requirement 'vigilance shall

Fig. 11 Sensor picture plus synthetic view (Source: Buro and Neujahr 2007)

be maintained by each person operating an aircraft so as to see and avoid other aircraft" (FAA 2004).

UASs do not naturally include such a feature. As there is no visual sense on board the UAV technical means, i.e. a suitable sensor or combination of sensors must be adopted to replace the eye of the pilot. The function demanded from UAVs is therefore called *sense and avoid*. It must provide an equivalent level of safety as the *see-and-avoid* capability of manned aircraft.

The relevant sensor(s) must be able to detect co-operative and non co-operative air traffic in the immediate vicinity of the own platform. Co-operative detection methods such as TCAS (Traffic Collision Avoidance System) and ADS-B (Automatic Dependent Surveillance-Broadcast) depend on the availability of data transmitting equipment on board the flying platform that can be received by other traffic. Non co-operative methods either rely on active (e.g. radar or laser) or passive (electro-optic or infrared cameras) sensors on board the platform. They are the only means to detect traffic which is not equipped with TACS or ADS-B, e.g. gliders, ultra-light aircraft or balloons, i.e. traffic not controlled by ATC (Air Traffic Control).

To detect a potential intruder, however, covers just the *sense* part of a *sense-and-avoid* function. The *avoid* part implies the initiation of an evasive manoeuvre in reasonable time. Depending on the distance and geometry of the conflict, the time for reaction can be very short, so that the commander of a UAV, considering potential latencies, might not be able to manually command the necessary manoeuvre. The logical consequence is to supply the UAV with the autonomy to independently carry out the necessary manoeuvre. This, however, means to take the UAV commander temporarily out of the control loop, which raises again the question who is going to assume responsibility should the manoeuvre fail, causing an accident.

Even if in certain cases automatic avoidance might become necessary, in the majority of situations there is enough time for the UAV commander to take manual action. To that end he needs a human–machine interface that enables him to act as

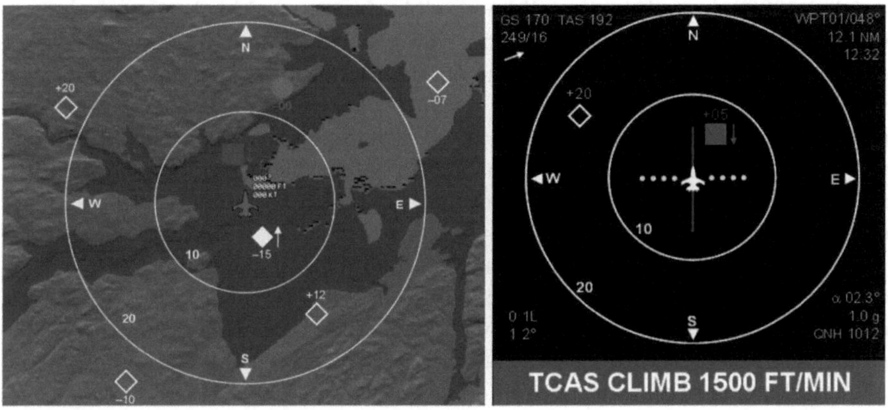

Fig. 12 Barracuda TCAS-HMI: Map Display (*left*) and command and control display (*right*) are designed in accordance with manned aircraft cockpits

required by the situation. In 2010, Cassidian successfully demonstrated TCAS in flight trials with its experimental UAV Barracuda. The relevant HMI is shown in Fig. 12. It includes indications on the Map Display (left) as well as on the Command and Control HMI (right) and is strongly oriented on the design of manned aircraft cockpits for maximum commonality. The UAV reaction to a Resolution Advisory (RA) was fully automatic but provided the operator with the option to manually intervene.

Activities to investigate technologies, methods and operational procedures are on-going since more than a decade. Partial functionalities have successfully been demonstrated, but a complete system, which has been certified by the authorities, does not exist yet. A potential reason might be the undefined responsibility during autonomous manoeuvring of the UAV.

3.5 *Future Prospects*

The design of future UAS ground stations depends on the operational capabilities required by the armed forces on the one hand and the technological options to put these requirements into practice on the other hand. Every novel technological feature will affect the design of the ground control station and its HMI.

The most frequently mentioned capability is higher platform autonomy. This is needed to absorb a potential shortage of UAV operators, particularly those with a pilot background. But autonomy is also required to expand the functions that can be carried out automatically thus enabling one operator to guide and control more than one UAV at any one time. Assuming that the UAV commander is only responsible for UAV guidance and not for payload operation, and provided that the UAVs hold a multitude of automated functions, a very preliminary estimation (Neujahr and

Reichel 2009) showed that it will be possible, from a workload point of view, to have two UAVs involved in a specific mission guided and coordinated by one commander. The operator still has the option to impact the operation and interact with the individual UAVs utilising a fairly conventional HMI that offers the necessary situational awareness, but at the same time it is clear that the analogy to manned aviation is lost. It never happened before that one pilot controlled two aircraft at the same time. Such a design, however, is certainly critical under certification aspects.

The next step in the same direction is the control of a larger number of UAVs by one operator, known as UAV swarm control, where the swarm operator, if at all, has only limited possibility to interact with individual UAVs but usually with the swarm as an entity via the assignment of duties. The details of the accomplishment of the task are defined by the swarm and managed internally. However, even such advanced UAVs require human supervision. As a result, further improvements of the UAV ground control station will be necessary as today they are inadequate for controlling and monitoring the progress of a swarm of UAVs (Walter et al. 2006). Even if the UAVs could be provided with enough intelligence to act as intended by the swarm commander, there are some unsolved issues with regard to the HMI design at the ground control station. According to Walter et al. (2006) the commander requires three things:

(a) Access to the control algorithm used by the swarm to enable either direct or indirect control,
(b) He must be able to monitor the swarms progress on a global level without getting caught up in the state of individual units, considering that there may be hundreds of UAVs in a swarm and
(c) He must be trained adequately to manage swarms.

From today's perspective such a system is, certainly not certifiable since the commander cannot instantly intervene should things go wrong, be it that he is not provided with detailed enough information only on swarm, not on an individual level or he is not able to quickly address individual UAVs. This prevents him from taking the role as the last safety instance.

For some military missions e.g. SEAD,[11] experts expect an operational benefit from a collaboration of manned and unmanned units. Regardless of whether the contributing UAVs are controlled from a ground station or from the cockpit of a manned platform (e.g. helicopter) a number of human factors issues have to be solved. Svenmarck et al. (2005) have identified the following critical items:

- Authority, i.e. who is in charge in a certain situation?
- Responsibility, i.e. who does what, when and how?
- Intentionality, i.e. is the intention of the UAV observable and understandable?

[11] Suppression of Enemy Air Defence: Activity that neutralises, destroys, or temporarily degrades surface based air defenses by destructive and/or disruptive means" (Joint Chiefs of Staff 2010).

- Trust, i.e. do the other members of the team trust that the UAV will take the right action in a certain situation, and does it perform the right action?

All of these issues need to be addressed for successful collaboration and coordination between manned and unmanned platforms.

The above mentioned trends are pointing towards a more remote future, but some changes are already becoming apparent with regard to the ground stations of current concepts, which are characterised by a one-to-one relationship commander-UAV:

- Standardisation of ground stations: today every supplier of a UAS designs and builds his own ground station, which, despite all standardisation efforts in assorted areas (e.g. data link), is largely incompatible with other UAVs. Ground stations should provide common functionality and for some basic functions a common *look and feel* for the operator. This is similar to the standardised layout of the *Basic-T* instrumentation in an aircraft cockpit, just to quote one example. The reasons for a standardised ground station are twofold: cost pressure and operability.
- Scalable control systems: it is imaginable that UAV control can be given (at least partly and temporarily and dedicated to a particular purpose) from the designated ground control station to an operator who is situated much closer to the spot of action than the actual UAV commander. The access to certain UAV functions should be possible utilising a small, portable HMI, such as a laptop or handheld computer.
- Growth of the number of systems, primarily sensors and communication systems on one platform: the same trend can be observed in manned aviation, resulting in an increased complexity of the ground control station which will only in part be compensated by a higher level of automation in these areas.

A further aspect is less related to the design of the UAS itself than to the way UASs are developed. Manned military aircraft are designed according to established systems engineering methods, associated with an iterative, top-down structuring of system requirements. This makes sure that the complex system is designed to meet specified user requirements and that all disciplines are included in the development process. The process takes a relatively long time but results in a mature product. In most cases, UASs in contrast seem to be the result of technology demonstration programmes, which were put to test in real military conflicts because the forces were hoping to profit from their deployment. The operational experience, in turn, continuously generated new requirements that had to be fulfilled very quickly by an upgrade of the UAS. In doing so, new functions and capabilities were introduced gradually, the main focus of which was in most cases on the aircraft systems, sensors, data link and armament, whereas only little priority was given to optimise the human–machine interface in the ground station. A big step towards more operational reliability and user satisfaction will be a UAS that utilises state-of-the-art HMI technology and is engineered according to established standards.

References

US Air Force. (2009). United States air force unmanned aircraft systems flight plan 2009–2047. http://www.govexec.com/pdfs/072309kp1.pdf. Accessed 31 July 2011.

Alberts, D. S., Garstka, J. J., Hayes, R. E., & Signori, D. A. (2001). Understanding information age warfare. Library of congress cataloging-in-publication data. http://www.dodccrp.org/files/Alberts_UIAW.pdf. Accessed 27 June 2011.

Asaf Degani and Michael Heymann (2002). Formal Verfication of Human-Automation Interaction

Ayton, M. (2011). Lockheed Martin F-35 lightning II. *Air International, 5*, 37–80.

Becker, S., Neujahr, H., Bapst, U., & Sandl, P. (2008). Holographic display – HOLDIS. In M. Grandt & A. Bauch (Eds.), *Beiträge der Ergonomie zur Mensch-system-integration* (pp. 319–324). Bonn: DGLR.

Billings, C. E. (1997). *Aviation automation: The search for a human-centered approach*. Mahwah: Lawrence Erlbaum.

Boucek, G., de Reus, A., Farrell, P., Goossens, A., Graeber, D., Kovacs, B., Langhorne, A., Reischel, K., Richardson, C., Roessingh, J., Smith, G., Svenmarck, P., Tvaryanas, M., & van Sijll, M. (2007). System of systems. In *Uninhabited military vehicles (UMVs): Human factors issues in augmenting the force. RTO-TR-HFM-078*. Neuilly-sur-Seine: Research & Technology Organisation. ISBN 978-92-837-0060-9.

Buro, T., & Neujahr, H. (2007). Einsatz der simulation bei der integration von Sensoren in unbemannte systeme. In M. Grandt & A. Bauch (Eds.), *Simulationsgestützte systeme*. Bonn: DGLR.

Carlson, B. J. (2001). *Past UAV program failures and implications for current UAV programs* (Air University Rep. No. 037/2001-2004). Maxwell Air Force Base: Air Command and Staff College.

Eichinger, A. (2011). Bewertung von Benutzerschnittstellen für Cockpits hochagiler Flugzeuge. Dissertation, Universität Regensburg.

Endsley, M. R. (1999). Situation awareness in aviation systems. In D. J. Garland, J. A. Wise, & V. D. Hopkins (Eds.), *Handbook of aviation human factors*. Mahwah: Lawrence Erlbaum.

Endsley, M. R., & Kiris, E. O. (1995). The out-of-the-loop performance problem and level of control in automation. *Human Factors, 37*(2), 381–394.

FAA. (2004). Title 14, Part 91 General operating and flight rules. Section 91.113(b). Federal Aviation Regulations.http://ecfr.gpoaccess.gov/cgi/t/text/text-idx?c=ecfr&sid=8a68a1e416cccadbf7bd2353c8cbd102&rgn=div8&view=text&node=14:2.0.1.3.10.2.4.7&idno=14. Accessed 31 July 2011.

Goodrich, M. A., & Boer, E. R. (2003). Model-based human-centered task automation: A case study in acc system design. *IEEE Transactions on Systems, Man and Cybernetics, Part A, Systems and Humans, 33*(3), 325–336.

Harris, W. C., Hancock, P. A., Arthur, E. J., & Caird, J. K. (1995). Performance, workload, and fatigue changes associated with automation. *International Journal of Aviation Psychology, 5*(2), 169–185.

Hoc, J. M. (2000). From human–machine integration to human–machine cooperation. *Ergonomics, 43*(7), 833–843.

Hollister, M. W. (Ed.), Adam, E. C., McRuer, D., Schmit, V., Simon, B., van de Graaff, R., Martin, W., Reinecke, M., Seifert, R., Swartz, W., Wünnenberg, H., & Urlings, P. (1986). Improved guidance and control automation at the man-machine interface. *AGARD-AR-228*. ISBN: 92-835-1537-4.

Joint Chiefs of Staff. (2010). Dictionary of military and associated terms US-Department of Defense. http://www.dtic.mil/doctrine/new_pubs/jp1_02.pdf. Accessed 16 July 2011.

Jukes, M. (2003). Aircraft display systems. Progress in astronautics and aeronautics 204.

Kaber, D. B., Riley, J. M., Tan, K. W., & Endsley, M. R. (2001). On the design of adaptive automation for complex systems. *International Journal of Cognitive Ergonomics, 5*(1), 37–57.

Kellerer, J. P. (2010). Untersuchung zur Auswahl von Eingabeelementen für Großflächendisplays in Flugzeugcockpits. Dissertation, TU Darmstadt.

Kellerer, J. P., Eichinger, A., Klingauf, U., & Sandl, P. (2008). Panoramic displays – Anzeige- und Bedienkonzept für die nächste generation von Flugzeugcockpits. In M. Grandt & A. Bauch (Eds.), *Beiträge der Ergonomie zur Mensch-System-Integration* (pp. 341–356). Bonn: DGLR.

Kornwachs, K. (1999). Bedingungen verantwortlichen Handelns. In K. P. Timpe & M. Rötting (Eds.), *Verantwortung und Führung in Mensch-Maschine-Systemen* (pp. 51–79). Sinzheim: proUniversitate.

Krah, M. (2011). UAS Heron 1 – Einsatz in Afghanistan. Strategie & Technik (Feb): 36–40.

La Franchi, P. (2007). UAVs come of age. Flight International (7–13.08.): pp. 20–36.

Lee, J. D. (2006). Human factors and ergonomics in automation design. In G. Salvendy (Ed.), *Handbook of human factors and ergonomics* (3rd ed.). New York: Wiley.

Manning, S. D., Rash, C. E., LeDuc, P. A., Noback, R. K., & McKeon, J. (2004). The role of human causal factors in U.S. Army Unmanned Aerial Vehicle Accidents. http://handle.dtic.mil/100.2/ADA421592. Accessed 31 July 2011.

Neujahr, H., & Reichel, K. (2009). C-Konzept für eine Sensor-to-Shooter Mission. AgUAV-R-000-M-0109.

Newman, R. L., & Ercoline, W. R. (2006). The basic T: The development of instrument panels. *Aviation Space and Environmental Medicine, 77*(3), 130 ff.

NIAG. (2004). Pre-feasibility study on UAV autonomous operations. NATO Industry Advisory Group.

Office of Secretary of Defense. (2005). Unmanned aircraft systems roadmap 2005–2030. http://www.fas.org/irp/program/collect/uav_roadmap2005.pdf. Accessed 31 July 2011.

Office of Secretary of Defense. (2007). Unmanned systems roadmap 2007–2032. http://www.fas.org/irp/program/collect/usroadmap2007.pdf. Accessed 31 July 2011.

Onken, R., & Schulte, A. (2010). *System-ergonomic design of cognitive automation*. Berlin: Springer.

Parson, H. M. (1985). Automation and the individual: Comprehensive and cooperative views. *Human Factors, 27*, 88–111.

Roscoe, S. N. (1997). The adolescence of aviation psychology. In S. M. Casey (Ed.), *Human factors history monograph series* (Vol. 1). Santa Monica: The Human Factors and Ergonomics Society.

Schulte, A., Meitinger, C., & Onken, R. (2009). Human factors in the guidance of uninhabitated vehicles: Oxomoron or tautology? The potential of cognitive and cooperative automation. *Cognition Technology and Work*. doi:10.1007/s10111-008-0123-2.

Sexton, G. A. (1988). Cockpit – Crew systems design and integration. In E. L. Wiener & D. C. Nagel (Eds.), *Human factors in aviation*. San Diego: Academic.

Sheridan, T. B. (2006). Supervisory control. In G. Salvendy (Ed.), *Handbook of human factors and ergonomics* (3rd ed.). New York: Wiley.

Sheridan, T. B., & Verplank, W. (1978). Human and computer control of undersea teleoperators. http://www.dtic.mil/cgi-bin/GetTRDoc?AD=ADA057655&Location=U2&doc=GetTRDoc.pdf. Accessed 11 July 2011.

Svenmarck, P., Lif, P., Jander, H., & Borgvall, J. (2005). Studies of manned-unmanned teaming using cognitive systems engineering: An interim report. http://www2.foi.se/rapp/foir1874.pdf. Accessed 11 July 2011.

Walter, B., Sannier, A., Reiners, D., & Oliver, J. (2006). UAV swarm control: Calculating digital pheromone fields with the GPU. *The Journal of Defense Modeling and Simulation, 3*(3), 167–176.

Wickens, C. D. (2000). The trade-off of design for routine and unexpected performance: Implications of situation awareness. In M. R. Endsley & D. J. Garland (Eds.), *Situation awareness analysis and measurement*. Mahwah: Lawrence Erlbaum.

Wickens, C. D., & Hollands, J. G. (2000). *Engineering psychology and human performance*. New Jersey: Prentice Hall.

Williams, K. W. (2006). Human factors implications of unmanned aircraft accidents: Flight-control problems. http://www.faa.gov/library/reports/medical/oamtechreports/2000s/media/200608.pdf. Accessed 19 July 2011.

Air Traffic Control

Jörg Bergner and Oliver Hassa

1 Introduction

Air traffic control (ATC) is a service provided by ground-based controllers who ensure the safe, orderly and expeditious flow of traffic on the airways and at airports. They control and supervise aircraft departing and landing, as well as during flight by instructing pilots to fly at assigned altitudes and on defined routes.

As described in DFS (2007, 2010b), air traffic control services perform the following tasks:

- Prevent collisions between aircraft in the air and on the manoeuvring areas of airports;
- Prevent collisions between aircraft and other vehicles as well as obstacles on the manoeuvring areas of airports;
- Handle air traffic in a safe, orderly and expeditious manner while avoiding unnecessary aircraft noise;
- Provide advice and information useful for the safe, orderly and expeditious conduct of flight;
- Inform the appropriate organisations if an aircraft requires assistance from, for example, search and rescue as well as to support such organisations.

In many countries, ATC services are provided throughout most of the airspace. They are available to all aircraft whether being private, military or commercial. In "controlled airspace", controllers are responsible for keeping aircraft "separated". This is in contrast to "uncontrolled airspace" where aircraft may fly without the use of the air traffic control system and pilots alone are responsible for staying away from each other. Depending on the type of flight and the class of airspace, ATC may issue instructions that pilots are required to follow, or merely provide flight

J. Bergner (✉) • O. Hassa
DFS Deutsche Flugsicherung GmbH, Langen, Germany
e-mail: joerg.bergner@dfs.de

M. Stein and P. Sandl (eds.), *Information Ergonomics*,
DOI 10.1007/978-3-642-25841-1_7, © Springer-Verlag Berlin Heidelberg 2012

information to assist pilots operating in the airspace. In all cases, however, the pilot in command has the final responsibility for the safety of the flight, and may deviate from ATC instructions in case of an emergency. The system in the United States, the United Kingdom and Germany are described in FAA (2010), ICAO (2001) and Safety Regulation Group (2010). Sudarshan (2003) explains the global air traffic management system. Good introductions to local air traffic control are given by Nolan (2004) for the United States of America (including a historic perspective) and Cock (2007) for the European system. The German air traffic management system is described in Bachmann (2005), Fecker (2001) and Mensen (2004). The historic development of ATC in Germany is explained by Pletschacher et al. (2003).

In Germany, airspace is divided into "flight information regions". These regions are assigned to radar control centres where they are subdivided into several areas of responsibility called "sectors". There are radar control centres for the lower airspace which are called "area control centres" (ACC) and for upper airspace which are called "upper area control centres" (UAC). The boundary between upper and lower airspace is usually at about 24,000 ft (7,500 m). ACCs and UACs have to ensure that aircraft are kept apart by internationally recognised distances called "separation minima". These are 5 NM (9.26 km) horizontally and 1,000 ft (300 m) vertically, see DFS (2007, 2010b).

While the controllers at both ACCs and UACs are mainly responsible for en-route sectors, ACCs may additionally include Approach Control (APP), see DFS (2006a,b). Approach controllers guide flights in the vicinity of airports, in a zone called the Terminal Manoeuvring Area (TMA). This area serves as a safe area for arriving and departing aircraft at an airport. FAA (2011a,b) describe the airspace structure and procedures of the United States.

In contrast to an ACC or a UAC, where the controllers depend solely on radars (Fig. 1); the controllers at an airport control tower rely on visual observation from the tower cabin as the primary means of controlling the airport and its immediate surroundings (Fig. 2).

Tower controllers work in the cab of the control tower high above the ground. As soon as the passengers have been seated and the aircraft is ready for take-off, the aircraft is under the control of tower controllers. The controllers inform the pilots about the departure procedures and the take-off position, and issue the take-off clearance.

At the same time, the controllers make sure that no other aircraft land or block the runway while the aircraft is preparing to take-off. To prevent bottlenecks on the runway due to a constant change between aircraft taking off and landing, the tower controllers are in permanent contact with the pilots and the approach controllers.

As soon as the aircraft has departed, approach controllers take over from the tower controllers and guide the aircraft onto one of the many airways in accordance with set procedures. Approach controllers ensure that prescribed separation is maintained between aircraft. In the TMA, the separation minima are 3 NM (5.4 km) horizontally and 1,000 ft (300 m) vertically.

Fig. 1 UAC Karlsruhe operation room, using the P1/VAFORIT system

Fig. 2 A tower controller in his tower cab

Area controllers assume control of the aircraft from the approach controllers and, depending on where the flight is headed, pass the control of the aircraft on to area controllers of UAC.

When the aircraft approaches the destination airport, area controllers resume control of the aircraft and pass it on the Approach Control. To prepare for landing, approach controllers arrange the aircraft along a virtual extended centreline of the runway before the tower controllers take over. The separation minima must be precisely observed even when the aircraft is on final approach.

Finally, tower controllers guide the aircraft to a safe landing on the runway and then direct it while taxiing along the taxiway to reach the apron.

2 Air Traffic Controllers' Tasks

Usually a team of two air traffic controllers, a radar and a planning controller, controls every sector. However, in times of low air traffic volume, sectors and working positions may be consolidated. The following sections describe their individual functions in more detail.

2.1 Radar Controller

Radar controllers monitor the radar display and issue instructions and clearances to pilots via radiotelephony. Within their area of responsibility, radar controllers vector aircraft to provide separation or navigational assistance. According to DFS (2007, 2010b) radar controllers perform the following tasks:

- Adjust equipment to ensure accurate radar representations or instruct maintenance personnel to initiate appropriate measures;
- Provide radar service to controlled aircraft;
- Identify aircraft and maintain identification;
- Radar vector aircraft to provide separation or navigational assistance;
- Issue clearances and instructions to ensure that radar separation minima are not infringed at any time;
- Document clearances, instructions and coordination results and update them, as appropriate (on Flight Progress Strips, systems or other);
- Monitor the progress of flights;
- Issue information to aircraft about unknown targets and adverse weather areas which are observed on the radar screen, if deemed necessary and permitted by the current workload situation; aircraft shall be vectored around such areas upon request;
- Comply with all applicable control procedures, taking noise abatement procedures into account;

- Provide flight information to pilots about things like weather and traffic;
- Apply non-radar separation in the event of a radar failure.

When controlling air traffic, radar controllers have seniority over the planning controller assigned to them. Clearances, instructions and coordination details have to be documented on Flight Progress Strips (FPS). If deemed necessary and permitted by the current workload situation, controllers can inform pilots about adverse weather areas or guide aircraft around such areas.

In the TMA, the tasks of radar controllers are further subdivided into the following functions:

- Pick-up controller
- Feeder controller
- Departure controller

Pick-up controllers take over control of aircraft from adjacent ACC sectors and assign specific headings, speeds and levels in accordance with the planned approach sequence. They are also responsible for the surveillance of crossing traffic within their area of responsibility.

Feeder controllers take over radar service for arriving aircraft from the pick-up controllers and determine the final approach sequence by directing aircraft onto the final approach. Depending on the current traffic and meteorological situation, they issue clearances for instrument or visual approaches. Before aircraft enter the control zone, the feeder controllers hand them over to the airport controllers.

Departure controllers provide radar service to aircraft which have just left the airspace under the responsibility of tower controllers and transfer the control to the appropriate area controller.

2.2 Planning Controller

Planning controllers support the radar controllers by coordinating and planning, for example, transferring responsibility to and from adjacent sectors or control units. They obtain and forward information and issue clearances to adjacent control units. Their duties also include planning the expected traffic flow and making proposals on how traffic should best be handled. According to DFS (2007, 2010b) planning controllers perform the following tasks:

- Obtain and forward information required for the orderly provision of air traffic control;
- Issue air traffic control clearances to adjacent control units;
- Perform radar hand-offs to/from adjacent sectors or control units;
- Prepare and maintain a traffic picture of the current traffic situation and, if appropriate, inform the responsible radar controller about possible infringements of separation minima;

- Analyze and plan the expected traffic flow and propose solutions for conflict-free traffic handling;
- Assist the responsible radar controller in establishing and maintaining separation during system/radar failures;
- Assist the responsible radar controller in the case of emergencies (for example, by reading the emergency checklist out loud);
- Document clearances, instructions and coordination results and update them, as appropriate (on flight progress strips or using systems or other methods).

3 The Controller Working Position (CWP)

Radar controllers maintain two complementary views of air traffic. The first is the radar screen, which lets them keep track of the current position of the aircraft in question. The second is a collection of flight progress strips, which allows them to organize and keep track of the aircraft as well as plan a strategy and record key decisions Mackay (1999). In general, there is no difference between the working positions used by radar or planning controllers, except for the presets of the communication system (Fig. 3). Both controllers use radiotelephony (r/t) to communicate with pilots and telephones to communicate with each other. In addition, the controller usually has several support systems, well-adapted to the specific information needed for this working position (Fig. 4). The components of the German standard system for ACC, the so-called P1/ATCAS, described in Bork

Fig. 3 Typical radar and planning controller working positions (P1/ATCAS Munich ACC)

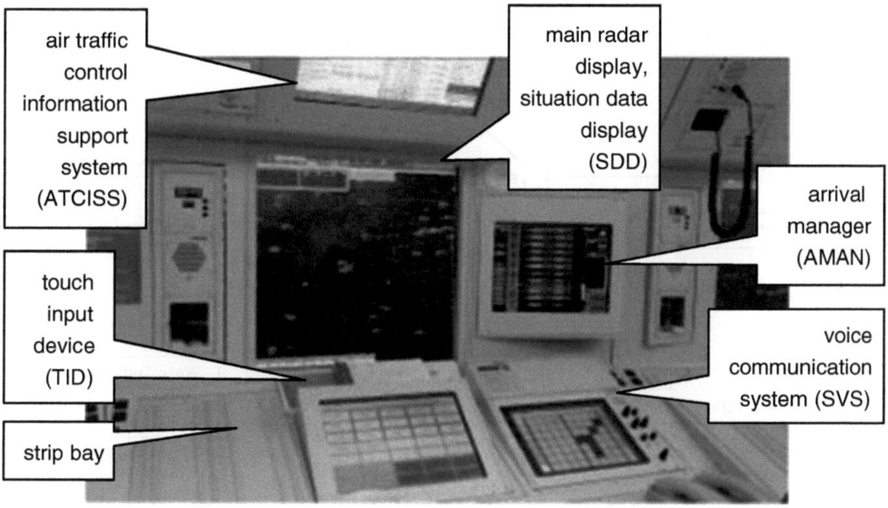

Fig. 4 An approach controller's working position

et al. (2007, 2010), DFS (2010a, e) will be used as an example of a typical working position in the next sections.

Hopkin (1995), ICAO (1993, 2000) describe the basic design of controller working positions in detail. Some of the European systems and their human–machine-interface are based on the work done by EUROCONTROL, see Eurocontrol (1998, 2006), Jackson and Pichancourt (1995), Jackson (2002) and Menuet and Stevns (1999). In the United States, the systems are based on the Federal Aviation Administration guidelines, see Ahlstrom and Longo (2001, 2003), Ahlstrom and Kudrick (2007) and Yuditsky et al. (2002). Kirwan et al. (1997) gives an overview about human factors input for the design of ATM systems.

3.1 Radar

The radar display provides a two-dimensional picture of aircraft moving in three-dimensional space. It displays synthesized information from primary and secondary surveillance radar. Each aircraft is represented by a symbol showing its current position, this is accompanied by a label that includes call-sign, current speed, flight level and often also a "history trail symbol" which shows recent positions as well as a predicted line of further flight (Figs. 5, 6).

Modern radar displays include additional categories of information relevant to the controller.These include things such as background maps which show airways, sector boundaries, location of navigation aids, runways, restricted airspace, geographical information, weather etc. Information based on computations like e.g. dynamic flight legs, Short Term Conflict Alerts (STCA), Medium Term Conflict Detection (MTCD), data from Arrival Managers (AMAN), etc. may also be depicted, see Bork et al. (2010).

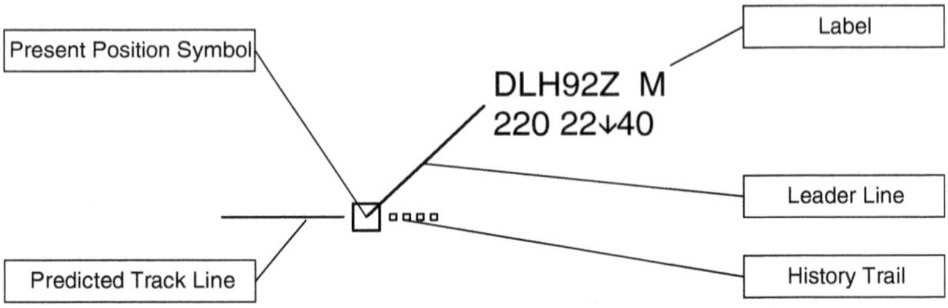

Fig. 5 Typical elements on the radar screen (P1/ATCAS, Munich APP)

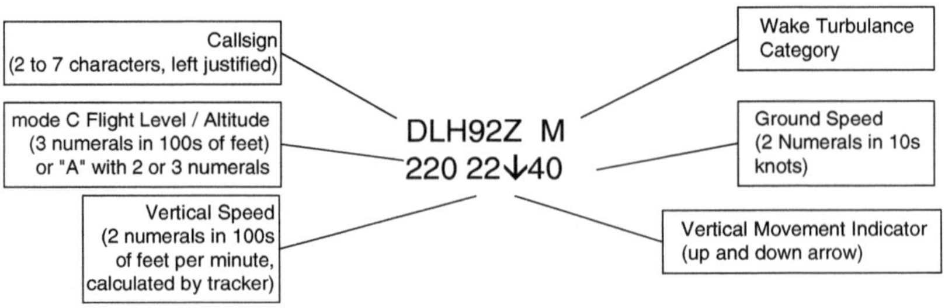

Fig. 6 Elements of the radar label (P1/ATCAS, Munich APP)

Controllers can modify the image, e.g. the magnification level. Additionally, they can set the screen to show only the flight levels relevant to their airspace sector or adjust the way the current position symbol is displayed, including the history trail, the predicted track line and the data shown in the label.

3.2 *Flight Progress Strips*

Most controllers in Germany still use pre-printed strips of paper, called flight progress strips, to keep track of the aircraft they are responsible for. Controllers can choose how to arrange the strips, which are covered in plastic holders, on a rack (strip board). The layout of the strips reflects the controller's personal view of the traffic. Additionally, controllers write on these strips using standard abbreviations and symbols so that they are easily understood by others, see DFS (1999, 2007, 2010b). Writing on them, arranging as well as organizing them on a strip board helps controllers to memorize information and build a mental picture of the air traffic. On flight strips one can find information about "the goals, intentions and plans of pilots and controllers and their recent actions" (Harper et al. 1989). They help the air traffic controller to be aware of the complex situation, see Endsley and Kiris (1995) and Endsley (1995).

In some centres, flight progress strips have been replaced by computer systems. Some just use an image of a paper strip including designators and handwritten strips on the computer display (DFS 2010d). More modern systems include functions that change the basic working method. These "strip-less" systems support the controller with various tools and techniques for tactical and strategic decision-making (DFS 2010c).

It took more than a decade to replace the paper strips, because actual strips of paper have many characteristics that a computer system cannot replicate. The term "handoff" which is used today to denote the computerized transfer of control of an aircraft from one sector to another originates from the physical hand-over of the flight progress strip from one controller to the other. Research on the augmentation and replacement of the paper strips include reducing the number of strips (Durso et al. 1998; Truitt et al. 2000a, b) or eliminating them completely (Albright et al. 1995), modifying strip holders (Médini and Mackay 1998), or creating electromechanical strips (Smith 1962; Viliez 1979), standard monitors (Dewitz et al. 1987; Fuchs et al. 1980), touch input devices (Mertz and Vinot 1999; Mertz et al. 2000), portable personal digital assistants (Doble 2003; Doble and Hansman 2003) and displays with pens (DFS 2010d). More information about automation of Air Traffic Management is given in Billings (1996, 1997), Fitts (1951), Wickens (1997, 1998).

3.2.1 Paper Strips

Paper flight strips are easy to use, effective at preventing failures and are easily adaptable to an individual controller's way of working. The main drawback of paper flight strips is that they are not connected to computerized tools. However, they are still used in modern Air Traffic Control as a quick way to record information about a flight, and they count as a legal record of the instructions that were issued, allowing others to see instantly what is happening and to pass this information to other controllers who go on to control the flight. See Hopkin (1995), Jackson (1989), Manning et al. (2003), Zingale et al. (1992, 1993) for further description of the use of paper based flight progress strips.

The strip is mounted in a plastic "strip holder" which colour in itself often has a meaning. A "strip board" filled with paper strips in their strip holders represents all flights in a particular sector of airspace or on an airport. In addition, the strip board has vertical rails that constrain the strips in several stacks, the so called "bays" (Fig. 7).

The position of the strip on the board is significant: approach and area controllers keep their strips in the landing sequence. The spatial orientation of the board represents a geographical area within an ATC-sector. Tower controllers might use the bays to make a distinction between aircraft on the ground, on the runway or in the air. Each bay might be further sub-divided using "designator strips" which communicate even more information by the position of a flight progress strip. Additional special strips can be used to indicate special airspace statuses, or to represent the presence of physical obstructions or vehicles on the ground.

Fig. 7 Flight progress strips in a strip-board with two bays

Additionally, a strip may be "flagged" that is placing at an angel in the bay so that one end sticks out. This is used to remind controllers of certain potential issues. This can be used either as a personal reminder or as a form of communication between the radar and planning controller.

There are many styles of progress strip layouts (Fig. 8), but minor differences aside, a strip contains at least:

- Aircraft identification (call sign);
- Aircraft-type as 4-letter ICAO designator (e.g. B744 for a Boeing 747–400);
- Flight level (assigned altitude);
- Departure and destination airport;
- At least one time in four figures (other times can be shortened to minutes only);
- Planned route;
- Other information may be added as required.

Further information is then added by controllers and assistants in various colours indicating the role of the person making the annotation (Fig. 9), see DFS (1999, 2006, 2010b). For analyses of the use of paper flight progress strips see Durso et al. (2004), Edwards et al. (1995), Vortac et al. (1992a,b).

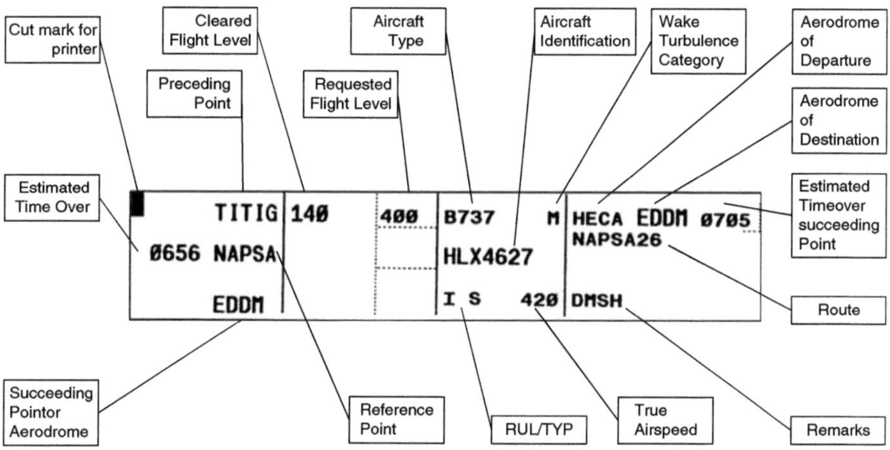

Fig. 8 Information shown on a flight progress strip (Munich APP)

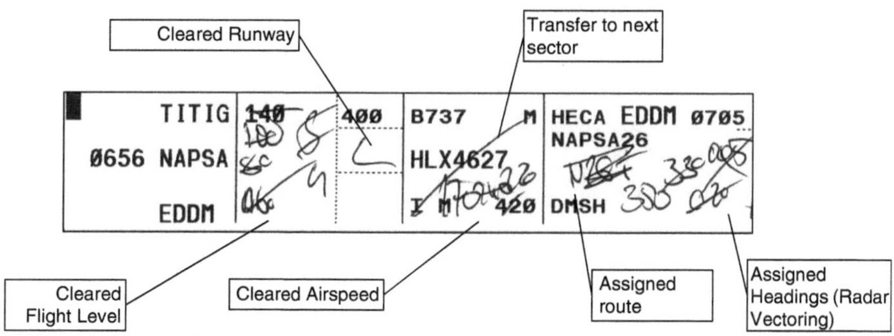

Fig. 9 Clearances documented on a flight progress strip (Munich APP)

3.2.2 Paperless-Strip-Systems (PSS) for Area Controllers

At every control position, the paperless-strip-system (DFS 2010d) has a flat panel display large enough to present images of flight strips in columns emulating the former flight strip bays for paper strips (Fig. 10), see Dewitz et al. (1987). The columns are divided into flight strip bays representing, for instance, flights approaching a particular point. Flight strips, currently printed on tag board provide vital flight information to users managing the traffic in their assigned airspaces. Presenting this information electronically eliminates the printing, manual distribution, and archiving of paper flight strips.

A new flight strip is presented to a controller at a special entry position on the screen. When the user selects the flight strip with a touch pen, it is automatically sorted into its correct bay and the new flight strip is temporarily highlighted to indicate its new position. The user may use the touch pen to drag a flight strip from one bay to another.

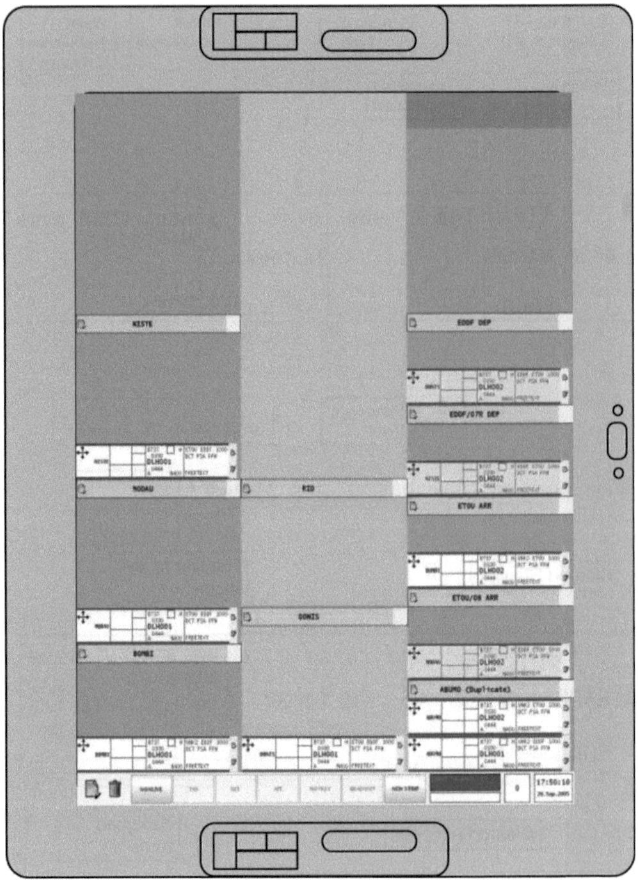

Fig. 10 Flight progress strips in PSS

The PSS can also present the user with pick-lists from which data, like a flight's assigned runway, can be selected via the touch pen (Fig. 11). The majority of user interactions with a flight strip, such as cleared flight levels transmitted to the pilots, are performed via a direct pick-list selection. Certain data that require a more flexible input range is entered into virtual keypads that pop up when needed. The system provides input syntax checking of the entered values as well as a validation of the values themselves. For example, the system reports if an entered cleared flight level poses a potential conflict with another flight path.

Controllers can write directly on the flight strips using the "scribble feature" and the touch pen. This feature is used primarily when the system is in a degraded state. Then blank flight strips are created and manually annotated in the same manner as the hand written updates made to the existing paper flight strips. As this data is electronically entered instead of written on paper, the data is available to both the

Fig. 11 Controller working with P1/ATCASS and PSS

system's processors and to other controllers. The system maintains an archive of each update that changes a value displayed on a flight strip.

3.2.3 Paperless-Strip-Systems for Tower Controllers

Since the primary means by which a tower controller keeps track of the current position of an aircraft is by looking out the window of the tower cab (Fig. 12), their working environment, and hence its information systems are different to that of radar controllers. This section describes some of the systems presently used by DFS, the German air navigation service provider at their control towers at the major international German airports.

Fig. 12 A tower controller working position at a major international airport

Tower Touch-Input-Device

To assist the tower controller at the two large German hub airports – namely Frankfurt and Munich – their working positions are equipped with a special front-end of the radar tracker (Bork et al. 2010). On the basis of correlated radar track information, the "Tower Touch-Input-Device" (Tower-TID) provides the tower controller with a timeline view of the arriving aircraft in the final approach pattern (Fig. 13).

Additionally, this concise representation assists the controller in keeping a record of the present status or ATC-clearance of the aircraft. For example, after the first radio contact and identification of the aircraft by the controller, the status "Initial Contact" is set for this particular aircraft by using the touch-screen. The colour of the aircraft label then changes according to its actual status or ATC-clearance.

Departure Coordination System

The Departure-Coordination-System (DEPCOS) is one of the central ATM-systems in the tower. It provides up-to-date flight plan information as well as all other necessary data about the status of departing aircraft in a monochrome, alphanumeric display (Fig. 14). The system comprises a custom-made keyboard (Fig. 15) that allows controllers to keep the sequence of the information that is displayed in the stack aligned with their plan of the departure sequence, see Siemens (1991).

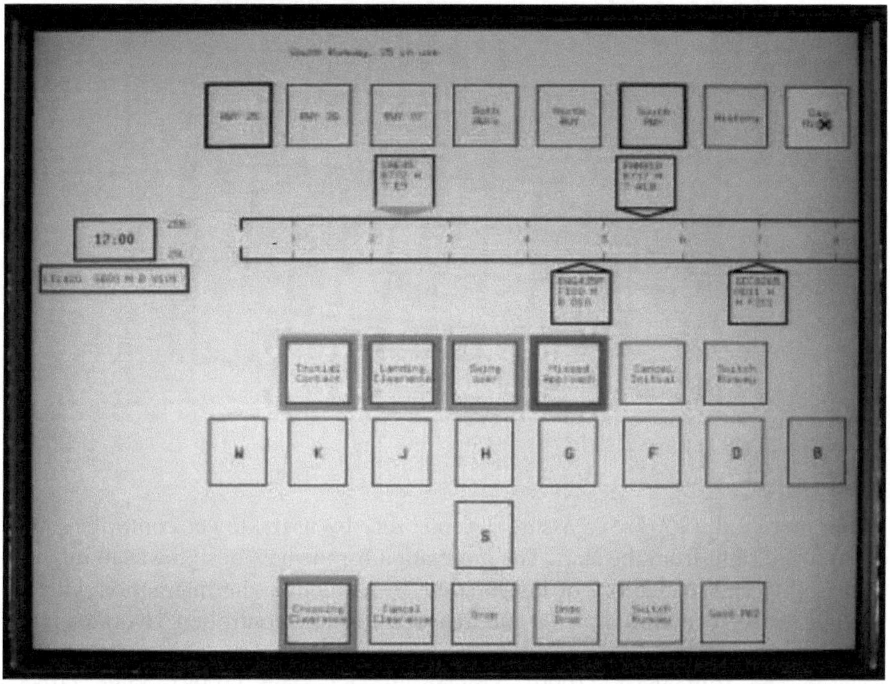

Fig. 13 Timeline view of the arriving aircraft on the Tower-TID Display

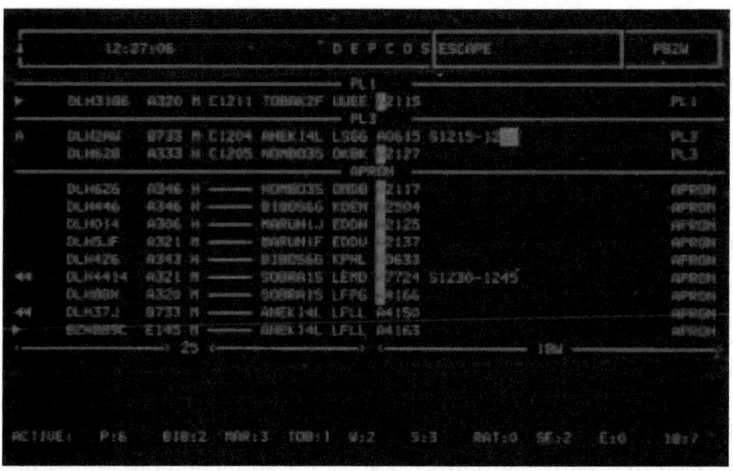

Fig. 14 Alphanumeric information about departing aircraft on the DEPCOS display

Integrated HMI for the Tower Controller

Considering the presently used systems for the arrival and departure information, an integrated HMI was developed using an inter-disciplinary approach combining operational, software, ergonomic and product design expertise, see

Fig. 15 DEPCOS keyboard

Bergner et al. (2009). To assure acceptability by users, tower controllers were involved right from the start. The motivation for the new design was to improve legibility and readability of information, to harmonize the interaction with the flight data processing systems and thus decrease the controllers' workload. For information about human performance in ATC see e.g. Eggemeier et al. (1985), Isaac and Ruitenberg (1999), Smolensky and Stein (1998), Stein (1985), Wickens (1992) and Wiener, E. L., & Nagel, D. C. (1989).

Depending on the importance of the information, it is displayed on different parts of the screen (Fig. 16). The controllers' attention is focused mainly on aircraft which will occupy the runway within the next few minutes. Information about these aircraft is placed at the centre of the display. Less attention needs to be paid to those aircraft that will occupy the runways at a later point in time and those that have not yet entered, or have left the controller's area of responsibility. That is why the information about these aircraft is placed in list form in the corners of the screen where it demands less of the controller's attention. As a result, there is also a physical separation of those objects on the screen that are important at this moment in time from those which are currently not important. This facilitates the overall visual orientation and weighting of the controllers while looking at the screen.

Following the concept of the Tower TID, a timeline principle was also used. The information associated with an aircraft is arranged in summarized form in labels. Their placement on the timeline corresponds to the time when an aircraft is expected to occupy the runway. Labels for all the aircraft whose take-off or landing is expected or planned within the next 6 min are displayed on the timeline where they move slowly towards the "present time". The differentiation between arrivals and departures is achieved not only by the information displayed but also by the background colour of the label. Departures are light gray, while arrivals are dark gray.

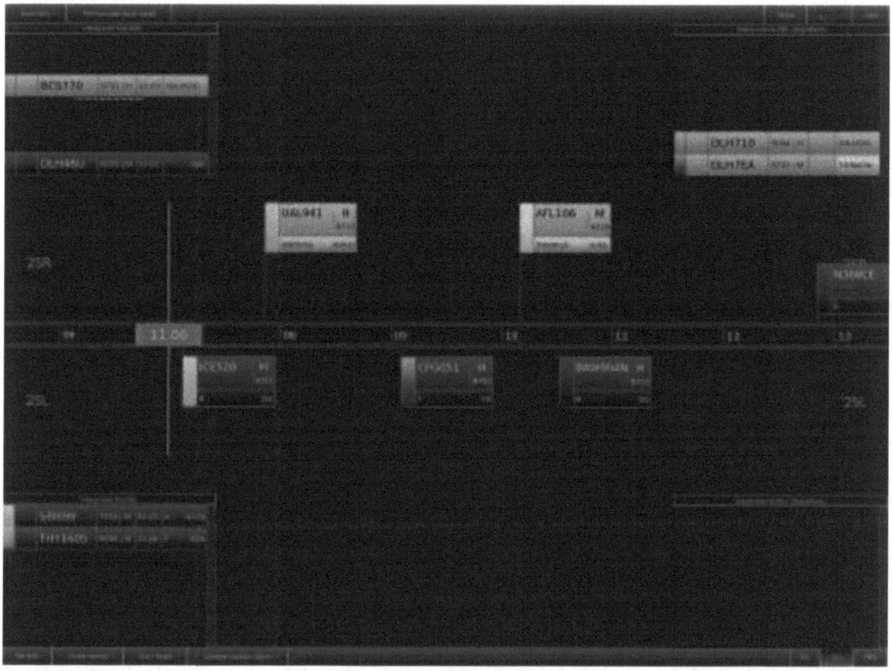

Fig. 16 Prototype of the new integrated HMI for the tower controller in Frankfurt/M

The timeline runs horizontally – mirroring the view from the tower towards the parallel runway system in Frankfurt. The present time is displayed on the mid-left of the screen. This corresponds to the approach direction of the aircraft for the runway-in-use 25, which is shown in the picture. The labels of aircraft for runway 25R are located above the timeline and those for runway 25 L below.

Arrival and departure labels for aircraft expected on the northern runway 25R first appear in the stack in the upper right corner of screen from where they move to the timeline. Depending on the designated apron-position north or south of the parallel runway, labels for landed aircraft move to either the top or bottom dark gray stack on the left side of the screen.

Labels for aircraft expected on the southern runway 25 L proceed likewise. After take-off, labels for departures move to the light gray stack in the upper left corner no matter from which runway they took off.

The amount of information in the stacks is reduced because at this point in time only certain limited pieces of information are important. Consequently, the data can be displayed in one line of the label. By stacking the labels, all the information of one type is located directly above each other. Thus the information can be absorbed at a glance more easily. The stacks are fitted with a scroll-bar to allow for the handling of more aircraft than a stack can display simultaneously. During normal operations, however, this feature is not supposed to be used very often.

The colour coding of the status information corresponds to that of the Tower TID. The presentation of colours takes into account the fact that they had to be easily recognizable, differentiated and non-dazzling for both day and night displays and different ambient light conditions. Since only about five colours can be easily distinguished at a glance, the application of colours was carefully chosen. Detail on the colour scheme can be found in Bergner et al. (2009).

The logical sequence of air traffic control clearances is implemented so that it requires only a click with the tip of the pen on the label head to progress through the successive status assignments. An editing function is available in the case of incorrect entry or if individual pieces of information are not correct. Using a context menu that is displayed by touching the label body, data can be edited.

3.2.4 Strip-Less-System for Area Controllers

The introduction of very advanced ATM Systems like the P1/VAFORIT System (DFS 2010c) and the operational changes planned will drastically change controller roles and working procedures.

The P1/VAFORIT System supports the controller with various tools and techniques for tactical and strategic decision-making. Certain tasks are performed system-automated by computers to increase capacity while controllers remain responsible for making decisions and thus ensuring a safe, orderly and uninterrupted flow of air-traffic.

The controllers are supported by an advanced Flight Data Processing System (FDPS) which includes software tools like conflict detection (Medium Conflict Detection, MTCD) and conformance monitoring which free controllers from routine tasks like coordination of flights. The Human-Machine-Interface (HMI) is designed to support the use of the automated system and the software tools. Paper control strips are substituted by a continuous trajectory based on electronic presentation of flight data.

The ATM system includes functions to enable the:

- Cooperative interaction between machine and controllers;
- Processing of more precise 4-dimensional trajectories;
- Support of automated coordination;
- Prediction and analysis of conflicts and monitoring of the flight progress;
- Elimination of paper control strips;
- Flexible distribution of tasks and data to controller working roles.

A CWP of the P1/VAFORIT system is shown in Fig. 17. It consists of:

- An Air Traffic Control Information Support System (ATCISS) monitor;
- Communication system components;
- The display components:

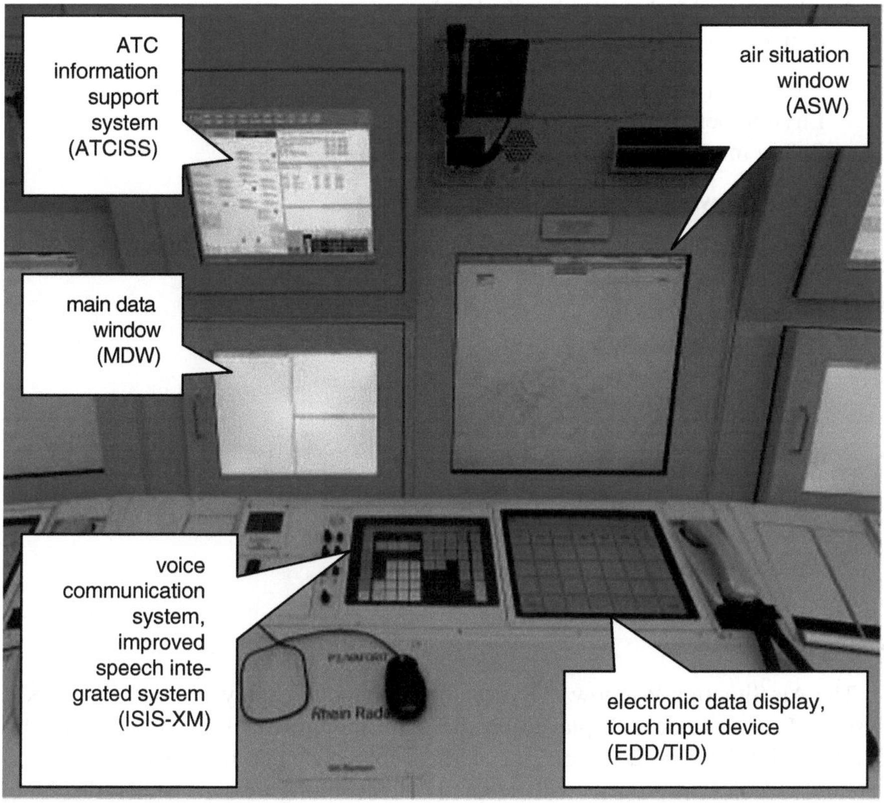

Fig. 17 Strip-less P1/VAFORIT CWP of a radar controller

- SDD (Synthetic Dynamic Display), displaying the ASW (Air Situation Window)
- MDW (Main Data Window) monitor
- EDD/TID (Electronic Data Display/Touch Input Device)

The purpose of the display system is to provide the system interface to the controllers. This interface presents radar data and flight plan data in graphs and tables. The interactive user interface, the Touch-Input-Device (TID) and the mouse are used by air traffic controller to input data in the P1/VAFORIT system.

The display consists of the following windows:

- Management Windows:
  - Preference set window
  - Check-in window
  - Configuration window
  - System mode window

- Air Situation Windows (radar displays)
- Information windows:
  - Flight plan window
  - Environmental window
  - Safety message window
- Tabular text windows:
  - Main data window
  - Pre-departure list
  - Inbound estimate list
  - Coordination windows
  - Lost linkage window
  - System error list
- Tool windows:
  - Stack manager window
  - Conflict risk display
  - Vertical trajectory window
  - Conflict information window
- Playback window

The Air Situation Window (ASW) is the main radar display of the P1/VAFORIT system. It is a window displayed on the main monitor of the CWP (28 in. TFT display with 2048 × 2048 pixels). This window cannot be closed, resized or rescaled. Upon start it opens with adapted default settings (range, centre, displayed maps and additional items and windows), upon login the preferences saved by the user take precedence.

Within this display (as well as within the monitor used for displaying the MDW) rules have been defined regarding the priority of display, appearance of windows and items and their attributes (resizing, rescaling, moving to other displays, etc.) which are adapted via a configuration tool. The controller interacts with the system via menus on the radar label and on items in the MDW. It displays the following menus: "Call sign" (C/S), "Flight Level" (FL), "Waypoint" (WPT), "Assigned Speed" (ASP) and "Assigned Heading" (AHD).

P1/VAFORIT enables the direct interaction with trajectories with the so called "Elastic Vector Function" and the Graphical Route Modification (GRM) tool. The "Elastic Vector" is a graphical tool to be used by the controller on the radar screen (ASW) for any of the following purposes:

- To input a direct route
- To input a closed track
- To measure the bearing and the distance between an object and the aircraft
- To input a direct route or a closed track in what-if sessions

The GRM is a graphical tool presenting a "rubber band" which is originated at an aircraft head symbol or a selected waypoint, following the same presentation rules as the Elastic Vector Function. It is used to modify or to probe modifications of a route (what-if session).

The MTCD function of P1/VAFORIT detects the existence of conflicts if aircraft trajectories are penetrating each other's safety volumes within a specified time (e.g. 20 min). When a MTCD conflict is detected between aircraft, a text message in an alert or warning colour in the alert/warning field of the data label will be displayed. In addition, the system is able to distinguish between those MTCD alerts or parts of MTCD alerts where flights are in vertical manoeuvres (vertical conflicts) and those that are only in levelled movement (lateral conflicts).

The Main Data Window (MDW) is the main tool for planning (and coordination) dialogues. This window provides aircraft information of the traffic planned. It displays active flight information for management purposes and coordination data when a coordination process has been triggered. The MDW gives access to the following actions:

- Flight level commands:
  - Change the cleared flight level and/or exit flight level
  - Assign a rate of climb/descent
  - Request an entry flight level from the previous sector
  - Define a level band as an entry or exit condition
- Holding command
- Speed command
- Heading commands
- Coordination dialogues: automatic and manual
- Flight plan markings: flight plan flag/unflag, mark/unmark individual elements, they are displayed and distributed to the controllers within the same sector

All information displayed in the "MDW" is continuously updated. Together with the MTCD and other support tools, it replaces the conventional, paper flight progress strips. It is permanently displayed at its maximum extent on a dedicated screen and consists of one window showing the:

- Main area,
- Departure area and
- Conflict area.

The work of the planning controller changed completely compared to the work with paper strips or a system with electronic flight strips (like PSS).

In addition to the interaction with the label on the SDD, many functions are also selectable by using the Electronic Data Display/Touch Input Device (EDD/TID). It is used to enter data into or request data from the system, and to manipulate data using an EDD/TID function set. Touching certain pushbuttons of the EDD/TID screen perform inputs on the EDD/TID. The EDD/TID provides the capability of executing the same actions as configured in the respective menus or tools or functions on the main screen (ASW) and on the MDW.

3.3 Support Systems

3.3.1 Arrival Manager (AMAN)

The AMAN is a support system for radar controllers, (Bork et al. 2010), (DFS 2010a), (DFS 2010e) and (Francke 1997). It assists controllers in ensuring safe and efficient planning, coordination and guidance of traffic from different inbound directions into the TMA of an airport. The system helps to reduce delay and distributes it equally over all inbound directions and follows the principle of "first come - first served", all while reducing the controller workload.

From the time an aircraft enters the Flight Information Region (FIR) until it has safely landed, the AMAN generates a sequence of messages. These are displayed on a set of special displays in form of a timeline. Whereas the AMAN information is presented to the approach controller on a dedicated monitor (Fig. 18), the area controllers use an AMAN window overlay on their radar screen (Fig. 19).

The AMAN uses adaptive planning, which takes the radar-track data and the flight plan data of each aircraft into account. Initially, an estimate is calculated as to

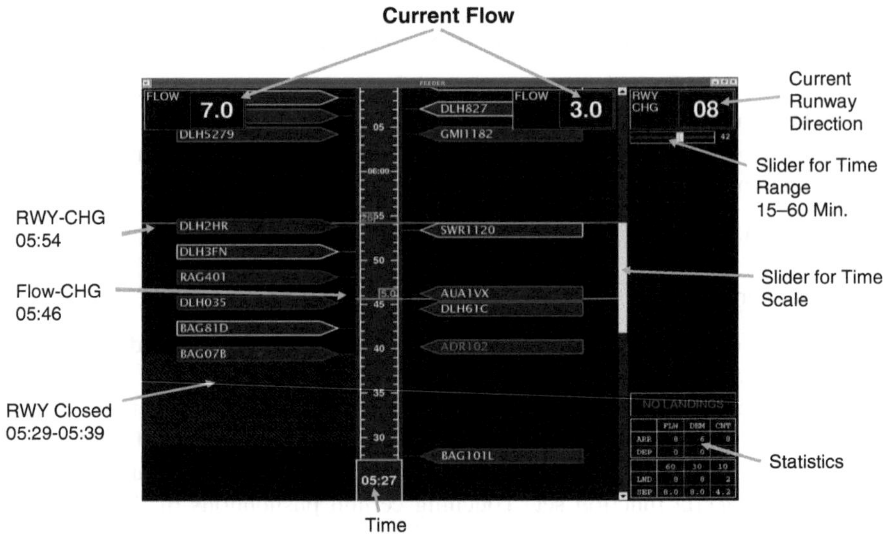

Fig. 18 AMAN display for the approach controller

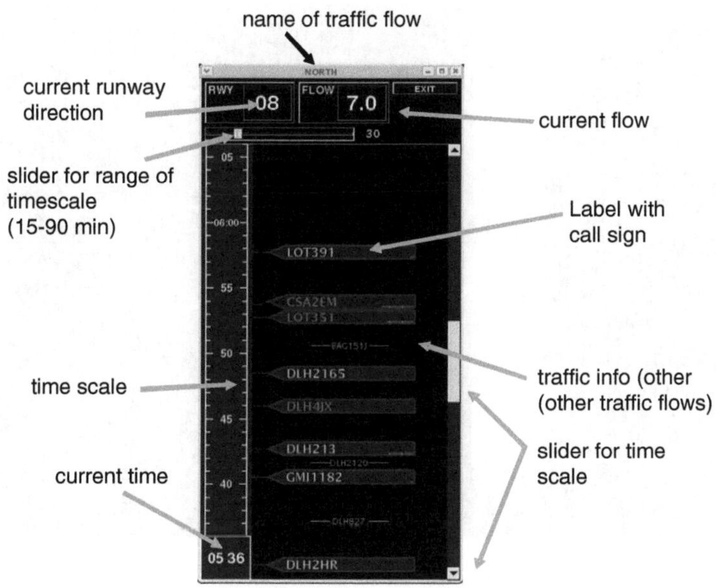

Fig. 19 AMAN display for the radar controller

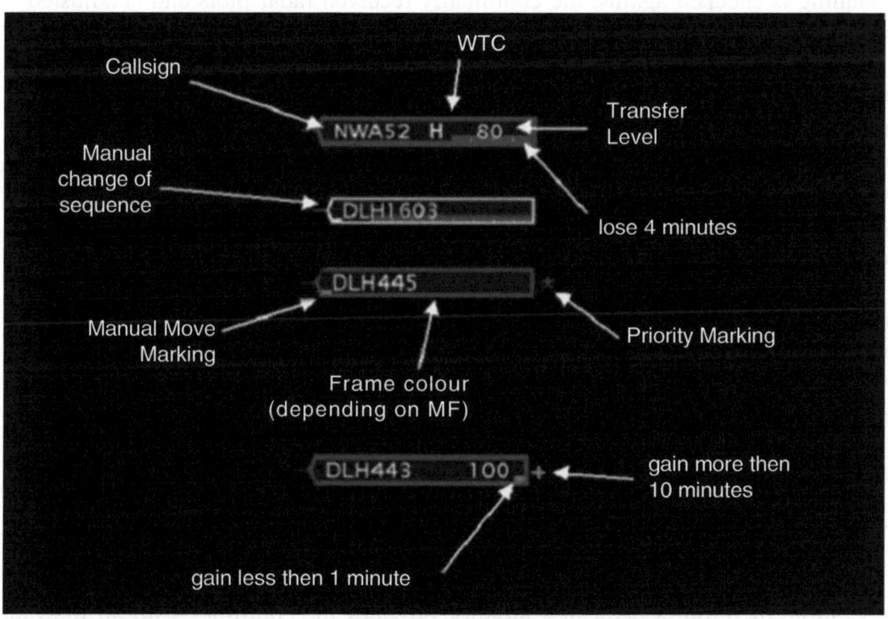

Fig. 20 AMAN label information for the radar controller

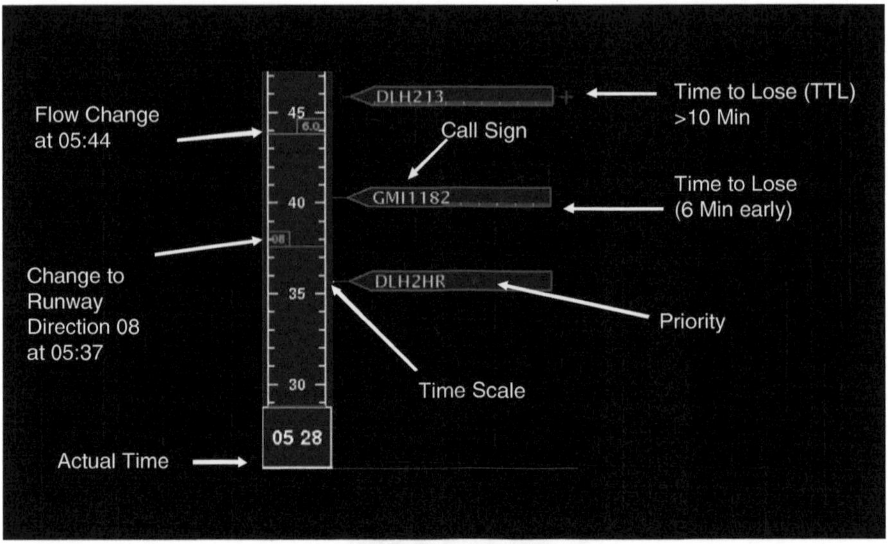

Fig. 21 Details of the AMAN display for the radar controller

the time when an aircraft will be over the metering fix, which it has to pass when entering the TMA, and the time when it will reach the runway. Subsequently, the planning is checked against the continually received radar plots and is adjusted if necessary, see DFS (2010e).

All planning is generally carried out on the basis of standard procedures. In case the system detects a deviation from these procedures, the planning is adjusted to reflect the actual situation. Thus controllers do not have to input anything in the case of a missed approach or a swing over of an aircraft on final from one runway to the other. Additionally, manual interventions by controllers such as a sequence change, a move, priority, runway change etc. are possible if it is necessary to adjust or improve the planned sequence (Figs. 20, 21).

The generation and updating of the sequence and the target times are done differently in ACC and APP. The aim of the planning for the controllers of the sectors adjacent to the TMA is to achieve the best possible planning stability, whereas the planning for the approach controllers first of all has to be highly adaptive to the actual guidance of the controllers.

3.3.2 Airport Weather Information Display

The meteorological conditions at the airport are of great importance for the safe operation, therefore a dedicated monitor provides the controller with all relevant weather parameters (Fig. 22).

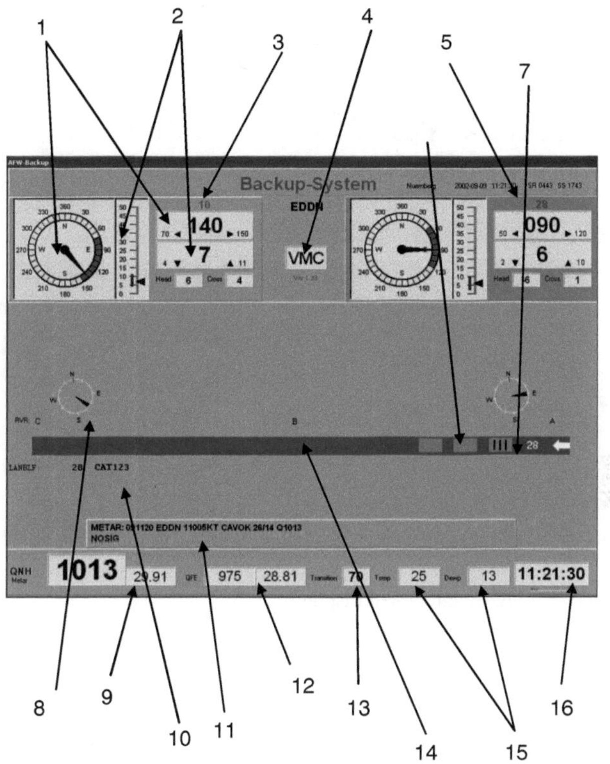

Fig. 22 Support system for airport weather information (DFS 2009)

Key:

1. Current wind direction with variation
2. Current wind speed with min and max values and head and cross wind components
3. Direction of runway to which the weather values relate
4. Current flight rules: Visual Meteorological Conditions (VMC) or Instrument Meteorological Conditions (IMC)
5. Sunrise, sunset
6. Current category of the instrument landing system
7. Active runway
8. Compass rose adjusted to runway orientation
9. QNH in hPa und mmHg (barometric pressure at mean sea level)
10. Information about ILS status (acronyms)
11. Current METAR ("Meteorological Aviation Routine Report")
12. QFE in hPa und mmHg (barometric pressure at airport ground level)
13. Transition level
14. Runway visual range (RVR)
15. Temperature and dew point in °C
16. Time

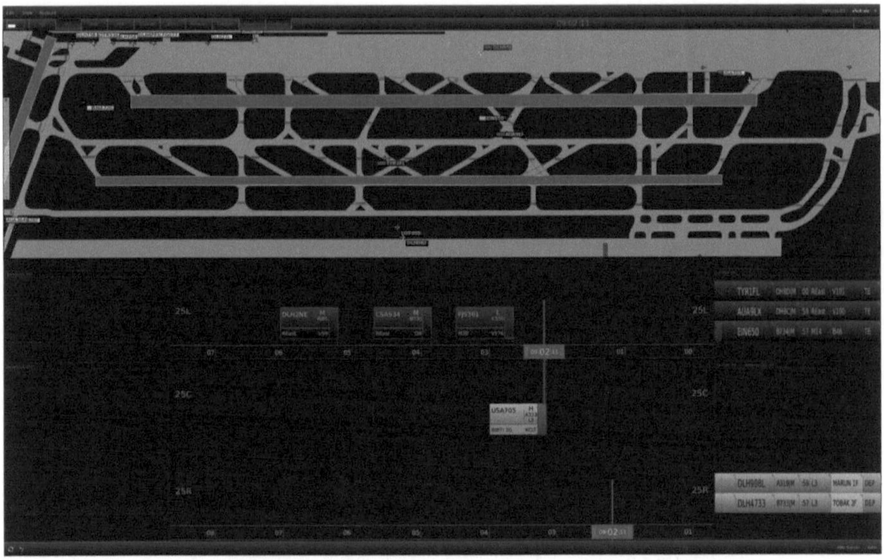

Fig. 23 Prototype of a highly integrated display

4 Outlook

Within the past two decades, advances in computer graphics – both hardware and software – have led to significant evolution of the human-computer-interface in the realm of air traffic control. Currently, considerable effort is being made to try to integrate related information that originates from separate sources in a concise and sensible way. An example is given in Fig. 23, which demonstrates the integration of arrival and departure planning-information along with ground radar and stop-bar switchboard functionality in a single display.

Additionally, fundamental research will be necessary to design HMIs that enable the controller to maintain situational awareness as higher levels of automation are going to be introduced to ATC in the future.

References

Ahlstrom, V., & Kudrick, B. (2007). *Human factors criteria for displays: A human factors design standard – Update of Chapter 5*. Atlantic City: Federal Aviation Administration William J. Hughes Technical Centre.

Ahlstrom, V., & Longo, K. (2001). *Computer–human interface guidelines: A revision to Chapter 8 of the human factors design guide*. Atlantic City: Federal Aviation Administration William J. Hughes Technical Centre.

Ahlstrom, V., & Longo, K. (2003). *Human factors design standard for acquisition of commercial off-the-shelf subsystems, non-developmental items, and developmental systems*. Atlantic City: Federal Aviation Administration William J. Hughes Technical Centre.
Albright, C. A., et al. (1995). Controlling traffic without flight progress strips: Compensation, workload, performance, and opinion. *Air Traffic Control Quarterly, 2*(3), 229–248.
Bachmann, P. (2005). *Flugsicherung in Deutschland*. Stuttgart: Motorbuch Verlag.
Bergner, J., König, C., Hofmann, T., & Ebert, H. (2009). An integrated arrival and departure display for the tower controller. *9th AIAA aviation technology, integration and operations conference (ATIO)*, Hilton Head. 21–23 Sept 2009.
Billings, C. E. (1996). *Human-centered aviation automation: Principles and guidelines*. Moffett Field: NASA Ames Research Centre.
Billings, C. E. (1997). *Aviation automation*. Mahwah: Lawrence Erlbaum.
Bork, O., et al. (2007). *P1/ATCAS trainings manual*. Langen: DFS Deutsche Flugsicherung GmbH.
Bork, O., et al. (2010). *ATM system guide lower airspace*. Langen: DFS Deutsche Flugsicherung GmbH.
Cook, A. (2007). *European air traffic management*. Aldershot: Ashgate.
Deutsche Flugsicherung. (1999). *Betriebsanordnung FMF-BC – Eintragungen auf den Kontrollstreifen für das ACC Frankfurt*. Langen: DFS Deutsche Flugsicherung GmbH.
Deutsche Flugsicherung. (2006a). *Betriebsanordnung 50/2006 Änderung der Sektor – und Rollenbezeichnungen*. München: DFS Deutsche Flugsicherung GmbH Niederlassung Süd Centre München.
Deutsche Flugsicherung. (2006b). *Betriebsanordnung 75/2006 operational order EBG APP, NORD, and SUED*. München: DFS Deutsche Flugsicherung GmbH Niederlassung Süd Centre München.
Deutsche Flugsicherung. (2007). *Betriebsanweisung Flugverkehrskontrolle*. Langen: DFS Deutsche Flugsicherung GmbH.
Deutsche Flugsicherung. (2009). *Benutzerhandbuch Betrieb L-ANBLF*. Langen: DFS Deutsche Flugsicherung GmbH.
Deutsche Flugsicherung. (2010a). Controller working position (CWP) system user manual (SUM) for P1/ATCAS (Air Traffic Control Automation System) release 2.8. Deutsche Flugsicherung GmbH Systemhaus/SoftwareEntwicklung Main ATS Components, Langen.
Deutsche Flugsicherung. (2010b). *Manual of operations air traffic services*. Langen: DFS Deutsche Flugsicherung GmbH.
Deutsche Flugsicherung. (2010c). *P1/VAFORIT user manual*. Karlsruhe: DFS Deutsche Flugsicherung GmbH, Upper area control centre.
Deutsche Flugsicherung. (2010d). Paperless strip system system user manual for P1/ATCAS release 2.8. Deutsche Flugsicherung GmbH Systemhaus/Software Entwicklung Main ATS Components, Langen.
Deutsche Flugsicherung. (2010e). System user manual Munich for AMAN release 3.1. Deutsche Flugsicherung GmbH Systemhaus/Software Entwicklung Centresysteme, Langen.
Dewitz, W., et al. (1987). *Grundsatzuntersuchungen zur Darstellung von Flugverlaufsdaten auf elektronischen Datensichtgeräten unter Nutzung des Experimental Work-position Simulators (EWS)*. Frankfurt/M: Bundesanstalt für Flugsicherung – Erprobungsstelle.
Doble, N. A. (2003). *Design and evaluation of a portable electronic flight progress strip system*. Cambridge: Massachusetts Institute of Technology.
Doble, N. A., & Hansman, R. J. (2003). *Preliminary design and evaluation of portable electronic flight progress strips*. Cambridge: Massachusetts Institute of Technology.
Durso, F. T., et al. (1998). *Reduced flight progress strips in en route ATC mixed environments*. Washington, DC: Office of Aviation Medicine.
Durso, F. T., et al. (2004). The use of flight progress strips while working live traffic: Frequencies, importance, and perceived benefits. Human factors. *The Journal of the Human Factors and Ergonomics, 46*(1), 32–49.

Edwards, M. B., et al. (1995). The role of flight progress strips in en route air traffic control: A time-series analysis. *International Journal of Human Computer Studies, 43*, 1–13.
Eggemeier, T. F., Shingledecker, C. A., & Crabtree, M. S. (1985). Workload measurement in system design and evaluation. In *Proceedings of the human factors society – 29th annual meeting*. Baltimore, Maryland September 29-October 3.
Endsley, M. R. (1995). Toward a theory of situation awareness in dynamic systems. *Human Factors, 37*(1), 32–64.
Endsley, M. R., & Kiris, E. O. (1995). The out-of-the-loop performance problem and level of control in automation. *Human Factors, 37*(2), 381–394.
Eurocontrol. (1998). *Eurocontrol EATCHIP phase III HMI catalogue*. Brüssel: Eurocontrol HQ.
Eurocontrol. (2006). *A human–machine interface for en route air traffic*. Brétigny-sur-Orge: Eurocontrol Experimental Centre.
Fecker, A. (2001). *Fluglotsen*. München: GeraMond.
Federal Aviation Administration. (2010). *Air traffic control*. Washington, DC: U.S. Department of Transportation.
Federal Aviation Administration. (2011a). *Aeronautical information manual*. Washington, DC: U.S. Department of Transportation.
Federal Aviation Administration. (2011b). *Aeronautical information publication United States of America*. Washington, DC: U.S. Department of Transportation.
Fitts, P. M. (1951). *Human engineering for an effective air-navigation and traffic-control system*. Washington, DC: National Research Council, Division of Anthropology and Psychology, Committee on Aviation Psychology.
Francke, S. (1997). 4D-Planer. DLR, Inst. für Flugführung, Braunschweig.
Fuchs, R., et al. (1980). *Einsatz von elektronischen Datensichtgeräten zur Ablösung von gedruckten Kontrollstreifen*. Frankfurt/M: Bundesanstalt für Flugsicherung – Erprobungsstelle.
Harper, R. R., Hughes, J. A., & Shapiro, D. Z. (1989). *The functionality of flight strips in ATC work*. Lancaster: Lancaster Sociotechnics Group, Department of Sociology, Lancaster University.
Hopkin, V. D. (1995). *Human factors in air traffic control*. London: Taylor and Francis.
ICAO. (1993). *Human factors digest N°8, human factors in air traffic control*. Montréal: International Civil Aviation Organization.
ICAO. (2000). *Human factors guidelines for air traffic management systems*. Montréal: International Civil Aviation Organization.
ICAO. (2001). *Air traffic management*. Montréal: International Civil Aviation Organization.
Isaac, A. R., & Ruitenberg, B. (1999). *Air traffic control: Human performance factors*. Aldershot: Ashgate.
Jackson, A. (1989). *The functionality of flight strips: Royal signals and radar establishment*. Great Malvern.
Jackson, A. (2002). *Core requirements for ATM working positions: An overview of the project activity*. Brétigny: Eurocontrol Experimental Centre.
Jackson, A., & Pichancourt, I. (1995). *A human–machine interface reference system for enroute air traffic control*. Brétigny: Eurocontrol Experimental Centre.
Kirwan, B. et al. (1997). Human factors in the ATM system design life cycle. In *FAA/Eurocontrol ATM R&D Seminar*. Paris: FAA/Eurocontrol. 16–20 June 1997.
Mackay, W. E. (1999). Is paper safer? The role of paper flight strips in air traffic control. *ACM Transactions on Computer–Human Interaction, 6*(4), 311–340.
Manning, C. A. et al. (2003). Age, flight strip usage preferences, and strip marking. In *Proceedings of the 12th international symposium on aviation psychology*, Dayton.
Médini, L., & Mackay, W. E. (1998). *An augmented stripboard for air traffic control*. Centre d'Études de la Navigation Aérienne, Orly Aérogares.
Mensen, H. (2004). *Moderne Flugsicherung*. Berlin: Springer.
Menuet, L., & Stevns, P. (1999). *Denmark-Sweden interface (DSI) human–machine interface specification: ACC/APP*. Brétigny: Eurocontrol Experimental Centre.

Mertz, C., Chatty, S., & Vinot, J. (2000). Pushing the limits of ATC user interface design beyond S&M interaction: The DigiStrips experience. *3rd USA/Europe air traffic management R&D seminar*, CENA, Neapel.
Mertz, C., & Vinot, J. (1999). Touch input screens and animations: More efficient and humanized computer interactions for ATC(O). *10th international symposium on aviation psychology*, CENA, Columbus.
Nolan, M. S. (2004). *Fundamentals of air traffic control*. Belmont: Thomson Brooks/Cole.
Pletschacher, P., Bockstahler, B., & Fischbach, W. (2003). *Eine Zeitreise*. Oberhaching: Aviatic Verlag GmbH.
Safety Regulation Group. (2010). *Manual of air traffic services. Civil aviation authority*. West Sussex: Gatwick Airport South.
Siemens, A. G. (1991). *DEPCOS handbuch Frankfurt/Main*. Frankfurt/M: Siemens AG.
Smith, M. A. (1962). Progress with SATCO. *Flight International, 2799*(82).
Smolensky, M. W., & Stein, E. S. (1998). *Human factors in air traffic control*. San Diego: Academic.
Stein, E. S. (1985). *Air traffic controller workload: An examination of workload probe*. Atlantic City: Federal Aviation Administration Technical Centre.
Sudarshan, H. V. (2003). *Seamless sky*. Aldershot: Ashgate.
Truitt, T. R., et al. (2000a). *Reduced posting and marking of flight progress strips for en route air traffic control*. Washington, DC: Office of Aviation Medicine.
Truitt, T. R., et al. (2000b). Test of an optional strip posting and marking procedure. *Air Traffic Control Quarterly, 8*(2), 131–154.
von Villiez, H. (1979). *Stand und Entwicklung der Automatisierung als Hilfsmittel zur Sicherung des Luftverkehrs*. Brüssel: Eurocontrol.
Vortac, O. U., et al. (1992a). *En route air traffic controllers' use of flight progress strips: A graph-theoretic analysis*. Washington, DC: Federal Aviation Administration.
Vortac, O. U., et al. (1992b). En route air traffic controllers' use of flight progress strips: A graph-theoretic analysis. *The International Journal of Aviation Psychology, 3*(4).
Wickens, C. D. (1992). *Engineering psychology and human performance*. New York: HarperCollins.
Wickens, C. D. (1997). *Flight to the future*. Washington, DC: National Academies Press.
Wickens, C. D. (1998). *The future of air traffic control*. Washington, DC: National Academies Press.
Wiener, E. L., & Nagel, D. C. (1989). *Human factors in aviation*. San Diego: Academic.
Yuditsky, T. et al. (2002). *Application of colour to reduce complexity in air traffic control*. Atlantic City: Federal Aviation Administration, William J. Hughes Technical Centre.
Zingale, C., Gromelski, S., & Stein, E. S. (1992). *Preliminary studies of planning and flight strip use as air traffic controller memory aids*. Atlantic City: Federal Aviation Administration Technical Centre.
Zingale, C., et al. (1993). *Influence of individual experience and flight strips influence of individual experience and flight strips*. Atlantic City: Federal Aviation Administration Technical Centre.

Railroad

Ulla Metzger and Jochen Vorderegger

1 Introduction

Due to European law (e.g. railway (interoperability) regulations 08/57, 04/49, and 07/59) and the abolition of border controls within Europe, new opportunities are arising for cross-border railway traffic. Trains do no longer have to stop at inner-European borders in the countries within the Schengen area for entry and customs formalities. While the time spent at the borders formerly was used for exchanging traction vehicles and drivers, railroad companies now try to save time and enter the other countries with the same vehicles and drivers. Therefore, the technical systems as well as the railroad operating rules have to be harmonized. This harmonization is a prerequisite for permitting European competitors free access to the railroad network. Currently, this is being worked on within the framework of the European Rail Traffic Management System (ERTMS). ERTMS consists of two parts: the European Train Control System (ETCS) and the Global System for Mobile Communications – Rail (GSM-R). This article is dealing with the ergonomic and human factors aspects of developing a standardized Driver Machine Interface (DMI) for ETCS.

2 What Does "Human Factors/Ergonomics" Mean?

The area of human factors/ergonomics (HF) deals with human factors influencing man–machine systems. Human factors experts are involved in designing the man–machine interface, with the goals of reducing human errors and their

U. Metzger (✉)
Deutsche Bahn AG TSS - Gesamtsystem Bahn Völckerstraße 5, 80939 München, Germany
e-mail: ulla.metzger@deutschebahn.com

J. Vorderegger
Inspectie Verkeer & Waterstaat, Rail en Wegvervoer (Rail and Road Transport) Utrecht, The Netherlands

M. Stein and P. Sandl (eds.), *Information Ergonomics*, 227
DOI 10.1007/978-3-642-25841-1_8, © Springer-Verlag Berlin Heidelberg 2012

consequences, as well as increasing productivity, safety, and comfort within the system (Kirwan 1994). These goals are reached through numerous approaches (e.g. Wickens et al. 1998):

- Design of equipment: e.g. display and control elements for the driver or the rail traffic controller (RTC), seats in the driver's cab.
- Design of physical environment conditions: e.g. lighting, temperature, noise.
- Design of tasks: specification of what humans have to do in the system and which role they play, irrespective of the equipment they use for accomplishing the tasks, e.g. by defining whether a task should be accomplished manually or automated.
- Training and selection of personnel: these are additional measures on the part of the human user of the system, which cannot replace, however, efforts to establish an optimum design of the technical side.

As not even highly automated systems can function completely without humans, human factors should be taken into account in all technical systems that involve humans in any form (Flurscheim 1983). What is essential in the area of human factors is the interdisciplinary approach, i.e. engineers, psychologists, and users of the corresponding systems (e.g. drivers, rail traffic controllers) usually work together on the design of human-centered (also: user-centered) systems (Norman and Draper 1986). User-centered or humane design means that the system's design should focus on the user's requirements, and not e.g. on simple technical feasibility or on the low costs of a solution (Billings 1997). Systems should be designed in a way so that they take advantage of the user's strengths (e.g. high flexibility in information processing and decision making) while taking into account human weaknesses and balancing them out with technical means if necessary. For example, high loads of information should be stored externally and easily accessible for the user (e.g. in the form of check lists the user can work off). Human factors, however, exceed this narrow approach by far. The team, too (e.g. RTC/driver/train personnel team), or the organization (e.g. safety culture) influences each user's behavior and consequently system safety and is thus being examined in terms of human factors.

3 Changes Due to ERTMS and Effects on the Driver

On the way towards a standardized European railway system, it is necessary to overcome the diversity of all the different operating regulations and signal systems that have been established within Europe over a long period of time (Pachl 2002). One result of different efforts is ERTMS, an operation control system of European railroad companies that is supposed to standardize train protection. As a control and train protection system, ETCS is one part of ERTMS. It represents a new and independent train protection system that allows existing safety systems to be integrated with the aid of so called specific transmission modules (STM). A train with ETCS equipment communicates with the trackside equipment on a defined "level" and in a defined system mode. Transition from one level to the next follows defined rules. One effect of ETCS is the reduction of localized signals and the display of signaling and other information in the cab. As of ETCS Level 2, only in-cab signaling will be used. Today, information provided by localized signals has already been enhanced by visual and auditive in-cab information. In some cases (e.g. the German Linienzugbeeinflussung (LZB) or the French Transmission Voie-Machine (TVM)), the technical system has been developed so far that light signals are switched off or are not used for signaling within the system anymore and only in-cab signaling is used instead. ETCS, however, provides in-cab displays that surpass the displays used so far in terms of number of readouts and functions (e.g. diagnosis of vehicle failure) by far. Accordingly, the DMI that will be used in the future will be as important as (or even more important than) the localized signal systems that are in use today.

Up to now, most drivers work in a hardware-based environment where fundamental information (e.g. status information of pressure gauges, indicator lamps) is displayed inside the driver's cab, and encoded information (e.g. signals and signs) outside the driver's cab alongside the track. In addition, the driver is provided with information about the route (e.g. the German Buchfahrplan (a collection of all information important for the driver), slow zones) and operational rules (e.g. traffic instructions/company directives). The driver needs to analyze and integrate the information from these sources and evaluate and group it regarding importance and urgency. He also needs to know when to access which information and then act accordingly. Therefore, drivers are trained and taught how to recall a high amount of information from their memory (e.g. about the route) – a process highly susceptible to human errors. With the implementation of ETCS Level 2 (partly also with Level 1), hardware-based display and control elements will mostly be replaced with software-based elements that exceed the amount of readouts of LZB for example. Similar to LZB, almost no information will be available anymore outside the driver's cab.

Today, the driver's job is based on a relatively complex set of skills and rules. A driver who does not know a certain route or system is not allowed to drive a traction vehicle and is also not capable of doing so. In the future, the driver will be more than a well-trained technician. He will represent the interface between the vehicle and the rest of the system. Besides driving the vehicle confidently and safely, he will need to be able to handle schedule information (also in an international context if necessary), diversions, energy-efficient driving, connections to other trains and much more. The tasks and responsibilities assigned to him will increase in terms of number and complexity (e.g. driving along unfamiliar routes abroad, foreign languages). The driver will still have to handle these situations on his/her own in the future as he/she will remain the only person in the cab. All in all, this means that the driver's tasks will be further diversified and that the DMI has to be designed accordingly. In order to meet the growing requirements, the driver needs adequate appliances like e.g. a display that recognizes the complexity of a task and provides the information needed in that situation while hiding other information.

These developments will change the strain on the driver as well as his/her performance and tasks (e.g. regarding trackside observation). The extent and the consequences of these changes on the driver and system safety have not been entirely examined yet.

Throughout the past 5 years, ETCS has been implemented in various countries within the framework of several projects. Feedback from these projects has provided important hints for the evaluation of the effects of ERTMS in the driver's cab and for the evaluation of the ETCS DMI. The results have been integrated into the most recent DMI specifications.

The following parts were changed with respect to previous versions (Metzger and Vorderegger 2004): First of all, the DMI specification has been updated with a complete description of the soft key technology; so now both touch and soft key technology are specified. The description of the main parts of the DMI (A, B, C and D) have been completely checked against the System Requirements (SRS) and modified where needed. The same has been done concerning symbols and sounds. The color and brake curve philosophy has been aligned with the ERTMS brake curve document. However, the most visible changes have been made on the data entry part. Not only has the document been reviewed against the SRS, but the whole dialog structure has been described in such detail that every window is exactly defined.

4 Development of a Standard ETCS DMI

4.1 Procedure

The goal formulated by the European Commission regarding the development of a European standard DMI (as a result of the interoperability directive) was to find a solution that can be manufactured by industry and used by the operators without any problems. The standard DMI should not only display ETCS, but also provide the possibility to integrate the European Integrated Radio Enhanced Network Environment (EIRENE), i.e. the interface for GSM-R as a means of communication. The development was to follow the approach of a user-centered design with the user's requirements being the focus of system development.

The ETCS DMI project started in 1991, with an orientation phase in the individual railway companies. This was followed by interviews and workshops with European experts in driving traction vehicles (1992) and the development of the design (1992–1993), a test, and a simulation (1992–1993). In 1993, simulator tests were conducted with about 130 drivers from different European countries. This means that the ETCS DMI is based on analyses and empirical research conducted with the aid of a simulator test (UCI/ERRI 1996) where all elements of interaction between the driver and ETCS were tested. That does not only include speed indication, but also symbols, tones and data entry. An analysis concerning current and future tasks of the driver and a study of ergonomic aspects completed the development process. The specification and design phase lasting from 1994 to 1996 was followed by a study of a combined EIRENE/ETCS interface (Ergonomic study of EIRENE Sub-Project Team) and additional studies regarding data entry (soft key, touch screen, and dialog structure), symbols and acoustic information (Ergonomic Data Entry Procedure for ERTMS/ETCS, Ergonomic studies of UIC). This approach ensures that the DMI can be understood by all European drivers and used by all European railway companies; Ergonomic Data Entry Procedure for ERTMS/ETCS 1998). Altogether, 240 drivers were given the opportunity of bringing in their views and comments on the DMI design. In 1998, a work group within the Comité Européen de Normalisation Electrotechnique (CENELEC) made up of European experts of the different railway companies and industry was assigned the task of transforming this first draft into a specification for the interface. In 2005, this process resulted in the CENELEC technical specification (TS 50459). The European Railway Agency (ERA), being the European system authority for ERTMS, refined this specification into a document that is fully compatible with the ETCS System Requirement Specification SRS 2.3.0.d. (ERA DMI Specification). This document is currently being worked on in order to adapt it to the Baseline 3 requirements of the ETCS System Requirement Specification. The basic concept is described in the following.

4.2 Theoretical Background

In order to give an insight into what information and elements are necessary on the display of a driver's cab, several models have been used (Norman and Draper 1986; Norman 1988):

(a) *Conceptual Model* or theory of driving a traction vehicle: What information does the driver need?
The conceptual model provides the most important variables and the correlation between the variables the driver needs (e.g. quantitative variables, tolerances and changes, destination display). It is based on physical reality. The driver's conceptual model about driving a traction vehicle includes variables like e.g. speed, distance, command, ETCS mode, ETCS level, route information, and the correlation between all these variables.
Besides these quantitative variables, there are also qualitative variables, e.g. knowledge of the route and schedule information. Other important aspects are the correlation between those variables (e.g. braking distance and speed), the amount of information at a certain point in time, and projection into the "future". Thereby, the strain on the driver caused by the system can either be too high or too low. This aspect has to be considered for the design of the ETCS DMI (e.g. through studies with the driver in a simulator).
(b) *Mental Model* or the driver's way of thinking: The mental model ensures that the information to be displayed is in line with the driver's way of thinking. If the mental model is not taken into consideration, there will be the danger of too much or too little information being provided on the display. The mental model

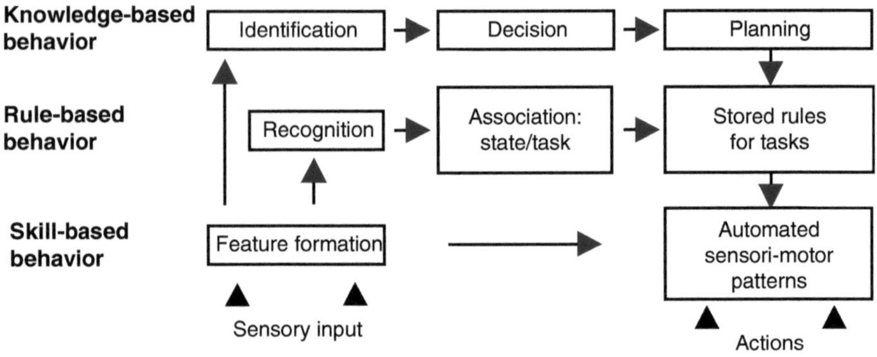

Fig. 1 Rasmussen's model

ensures that the conceptual information is also applicable for the driver. A high level of congruence between the displayed information and the user's requirements and expectations leads to high performance (i.e. short reaction times and only few errors). Cognitive ergonomics show how mental tasks are accomplished and how a high level of congruence can be achieved. Cognitive ergonomist Rasmussen (1986) distinguishes between the following behavior patterns of the operator (see Fig. 1):

Skill-based behavior: routine reactions or automated reactions to a stimulus, e.g. stopping at a stop signal, or using the acquired knowledge of the route. Building up skill-based behavior involves a lot of studying and training in the beginning, but with a lot of practice, this behavior occurs almost automatically.

Rule-based behavior: series of actions that are neither routine nor automatic, although they occur frequently, like e.g. passing a stop signal on command or estimating where to stop the train at the train station according to the length of the train.

Knowledge-based behavior: finding solutions for "new" problems, e.g. driving a new route or correcting a technical problem. Driving a traction vehicle on the level of knowledge-based behavior requires a considerable amount of experience, training, and knowledge of the system.

Each of these behavior patterns should be considered for the interface design. Therefore, a good balance of the conceptual and mental models is required. Important aspects for the mental model are, for example, choosing between absolute or relative values on the display (50 km/h, or too fast/slow), points or limits (0 km/h, or red), signals or speed (yellow-green, or 60 km/h), abstract or realistic presentation (e.g. the subway network presented as a pocket map compared to the real situation in the city). One important question is how to prevent skill degradation due to automation and how to design a system in a way so that it is easy to operate by both beginners and experts. Driver training, too, needs to be adapted to the modified requirements. Furthermore, it is necessary to consider that information which urgently needs to be displayed today could differ completely from what needs to be displayed in the future system. It is not necessary, for example, to display information (about the route) at present as route knowledge acquisition forms a large part of the driver's training. In the future, however, route information displayed in the cab will be indispensable as trackside signals will no longer be available.

ETCS imposes new requirements on the driver and his/her training. In the future, for example, the drivers will also have to drive on routes they are not (too) familiar with. This means that the driver will have to act on the knowledge-based level. As ETCS is capable, however, of supporting the driver by presenting route information on the display, the complexity of this task can be reduced to the level of skill-based behavior. Another example would be the task of correcting malfunctions within the technical system. Without support, this calls for knowledge-based behavior. But if the readout on the display provides a diagnosis or even instructions for the driver on how to solve the problem, the complexity of the required behavior is reduced to the skill-based level. This means that the driver will have to apply more complex (knowledge-based) behavior patterns in the future system which, however, can be reduced to skill-based behavior patterns with the system designed correspondingly.

(c) *Perceptual Model* The perceptual model explains how the system's design can meet the requirements arising from the features of the human senses. Important points here are e.g. human physical features (e.g. which wavelength of light can be perceived by the eye, which audio frequencies by the ear), the question of how to handle modifications in the display layout (e.g. additional presentation of a tone), or how glancing outside the cab can influence the perception of new information on the display inside.

It is important to consider and analyze each of these models because, taken as a whole, they represent the entire interface with the human operator. The better the comprehension, the better is the driver's performance and the lower the probability of errors. The target is to develop an optimum DMI based on (a) experiences with the current system, (b) knowledge about the future system and its requirements, and (c) the current knowledge about human factors.

4.3 Solution Description

Starting out from the System Requirement Specification (SRS) and the Functional Requirement Specification FRS, a CENELEC work group has created a document that describes the DMI for ETCS and EIRENE (ERTMS). As mentioned already, the ERA has transformed the ETCS part of the CENELEC TS into a DMI specification that forms part of the SRS.

The ETCS DMI represents a uniform philosophy and design approach for displaying the most important parameters. This is to ensure that the drivers can fulfill their tasks on a high performance level and that the risk of human errors is minimized. The document defines how ETCS information is displayed, how data is entered, what the symbols look like and what acoustic information shall sound like. It has been acknowledged that the DMI specification should not only describe the touch screen solution for the interface, but also the interface variant with soft keys. One important requirement was that no additional driver training is needed in order to be able to handle both interface solutions.

The ETCS DMI consists of six main fields that correspond with the driver's main tasks. Figure 2a, b show how the individual fields are used in the touch screen and the soft key variant.

(a) Monitored brake curve information (distance to target indication, prediction of the exact geographical position where the train will come to a standstill). This information is only displayed when the vehicle is in the brake curve. In Fig. 3, the distance to target is displayed on the left.
(b) Speed control (information about brake curve and speed control). This information is always displayed when the vehicle is moving. When the train is at standstill, this field can be used for other functions (e.g. data entry). Figure 3 shows an example of speed indication. Part of B is used for displaying announcements and information about the current mode.
(c) Additional driving information (e.g. ETCS level and information about contemporary train protection systems (NTC (National Train Control))).
(d) Planning of future events. An example of the forecast on route information relevant for a particular train (within the permitted driving route) is shown in Fig. 4. On the first forecast level, it is possible to present e.g. a speed profile matching the train with corresponding announcements, like tunnels, crossings, or changes in voltage, commands, or information on when the brake curve starts.

The forecast is to ensure that ETCS provides the driver with the same informa tion he has in the current system, without being provided with trackside information and information referring to the ETCS driving task under ETCS level 2. Instead, the forecast in the driver's cab provides the driver with route and schedule information obtained via balises and radio communication. The forecast under ETCS shows information that is dependent on the situation and only refers to that particular train. This means that route and schedule information is filtered and only information relevant for the corresponding train is displayed. The forecast also helps the driver to integrate and sort out information. For example, information on system safety (like the permitted maximum speed) is displayed separately from information that is not relevant for safety (e.g. the schedule), and information dealing with future events separately from information that is of immediate relevance. Separation of information helps the driver to focus on the correct part of the display and thus on the information with the highest priority. Due to this clear task-oriented display philosophy, the driver has more capacity left for other tasks, e.g. if he should have to react to unpredictable events in case of emergency. Thus, a contribution is being made to improve the overall systems safety.

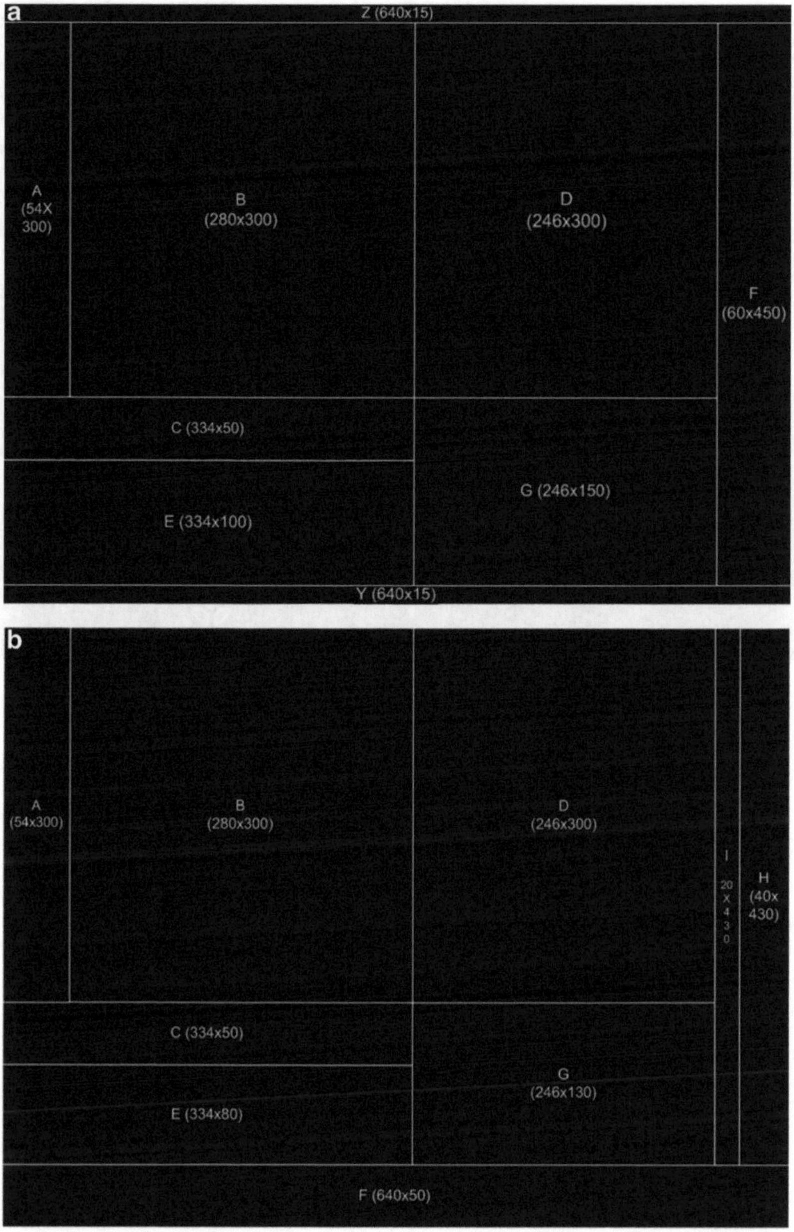

Fig. 2 (**a**) DMI fields as displayed in the touch screen solution. (**b**) DMI fields as displayed in the soft key solution

(e) Monitoring of the technical systems (alarms and status display of technical systems (e.g. door control, NTC, EIRENE information, time)). This applies to both ETCS and non-ETCS systems. There is the danger of every new system in

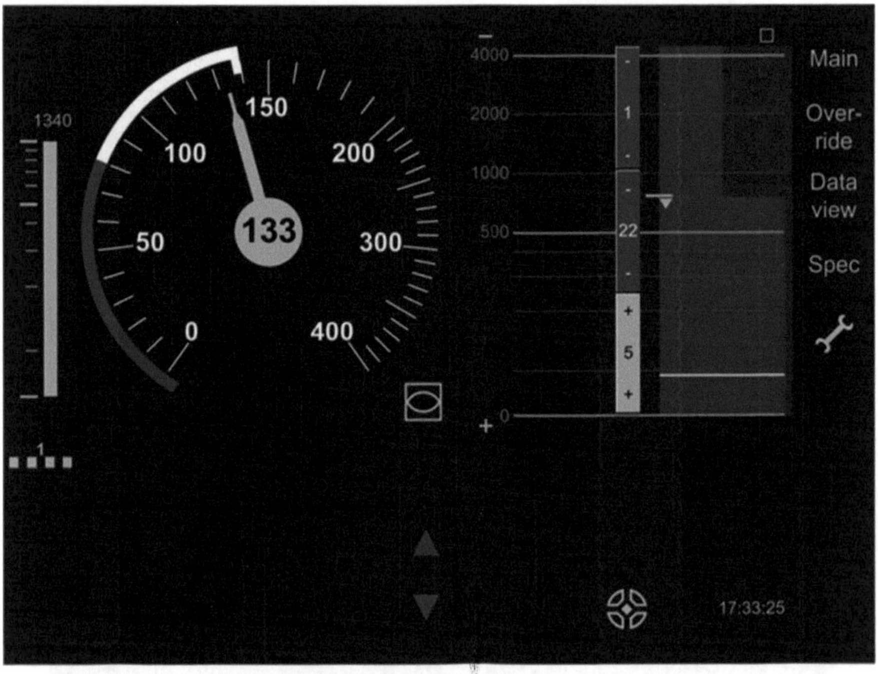

Fig. 3 Distance to target and speed indication

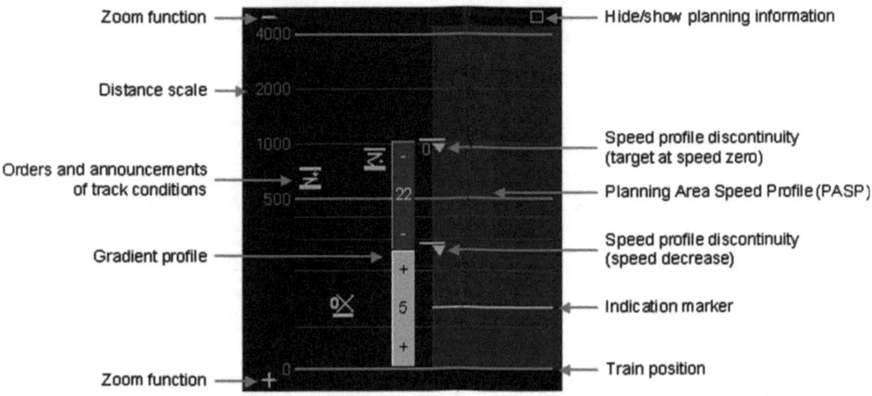

Fig. 4 Example of the forecast display

the driver's cab being equipped with its own interface without being adequately integrated with the other instruments. The ERTMS display tries to integrate all driving functions (e.g. speed control, data entry, EIRENE, door control, pantograph). For technical diagnoses and detailed information about braking and traction, separate displays will still be necessary. Although the DMI provides a

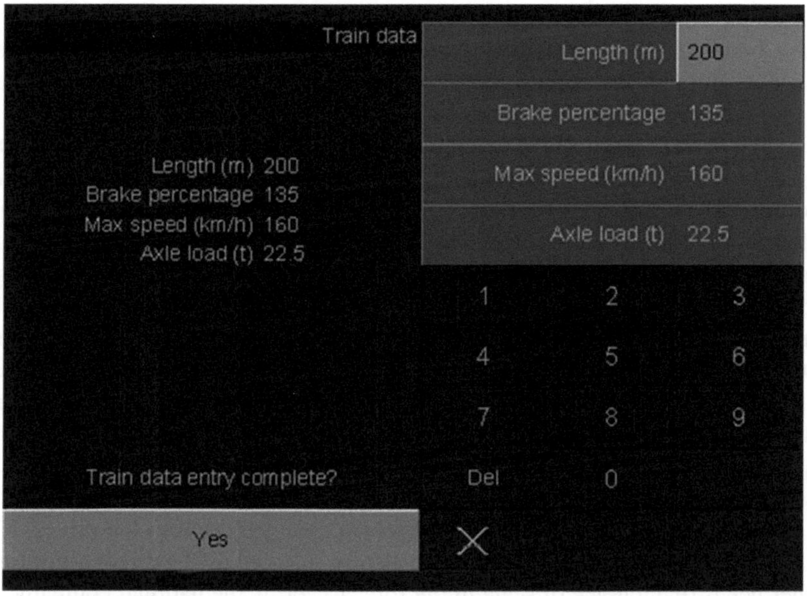

Fig. 5 Data entry window

lot of information, it has to be pointed out that a large part of this information is available in the current system already. Processing high loads of information imposes a lot of strain on the driver which can be reduced in ETCS (as it is already done in some national systems) through the display of information that is presented together, integrated and with certain techniques of attention control (e.g. hiding of unimportant information, prioritization).

(f) Data entry by the driver (entry of various train- and driver-related data). Figure 5 shows an example of the data entry window. The ERA DMI document provides a detailed description and specification of the data entry procedure in order to ensure that all ETCS DMI in Europe use the same dialog structure.

4.4 *Support for the Driver*

New technologies make it possible to provide the user with a large variety of information. The danger arises for the user to be overloaded. On the other hand, new displays offer the possibility of supporting the user in analyzing and integrating information, and of implementing new technologies to help the driver focus on relevant information. The latter has been implemented in the ETCS DMI through the application of various methods. *Acoustic signals* are used in the DMI in order to direct the driver's attention from the route outside towards new information on the display inside. Furthermore, very important visual information is enhanced by acoustic information. When the driver is on the verge of an automatic train stop or in the brake curve, information that refers directly to braking is made more noticeable through the application of symbols with a flashing frame. This means that the system provides the driver with information according to its priority. The driver first has to reduce speed before other information is displayed with the same degree of explicitness. Furthermore, the driver can adjust the forecast scale according to his/her own needs, i.e. the driver can choose a forecast of up to 32 km, or have only shorter distances displayed. Information outside the chosen range will thus not be displayed and does not burden the driver in his/her process of information processing. Nevertheless, the information is still available. Furthermore, there is a lot of information in the system which is available but rarely needed by the driver (e.g. in case of malfunctions or for testing purposes). Such data are available on another level and are only displayed when requested by the driver.

In this way, the system takes over part of the information integration, evaluation and presentation depending on importance and urgency in order to relieve the driver. The system can decide, for example, when information about the route or the schedule becomes relevant and will (only) then present it to the driver. However, the information to be displayed can also be selected by the driver according to his/her needs (e.g. choice of scale in the forecast) or by the national company. The information displayed in the forecast may, for example, depend on what information is being provided by the national company via data transfer.

The aim behind these methods of attention control is to fully utilize the possibilities and chances provided by ERTMS without exposing the driver to the risk of an information overload. While some national train protection systems provide drivers with relatively little information compared to ETCS and relatively little support for accomplishing their tasks, ERTMS will provide the driver with more information as well as more support and guidance from the system.

5 Summary and Future Prospects

In summary, it can be stated that ERTMS will cause a shift.

- From national to international traffic;
- From stops at the border to cross-border traffic;
- From hardware- to software-based systems;
- From external information display to internal information display;
- From non-integrated data display to integrated information display;
- From a relatively narrowly defined set of tasks to a more complex and comprehensive set of tasks for the driver;
- From tasks without system support for the driver's information processing to information processing supported by the system.

The effects of these changes on the driver have not been entirely examined yet. Eventually, it would be important to thoroughly reevaluate the specifications, which have further developed since the last tests, in simulator and field tests with drivers to ensure that the solution is useful and, even more important, easy to operate. That also includes collecting objective data on the performance of the driver and the stress experienced by the driver (e.g. reaction times, error rates, eye glance movements) during interaction with such a system. An eye-glance movement study of drivers interacting with ETCS has only just been completed at the simulator of the Deutsche Bahn AG.

The implementation of ERTMS is an opportunity for the railway companies to enter a new era with modern technology and human factors knowledge, and to bridge, at the same time, historically grown differences with a good solution. If all railway companies follow the DMI specifications, safety will be enhanced, training requirements for international traffic will be reduced, communication will be easier, and costs can be reduced on a long-term basis.

References

Billings, C. E. (1997). *Aviation automation: The search for a human-centered approach*. Mahwah: Lawrence Erlbaum.

CENELEC CLC TS 50459 – part 1 to 6; ERTMS – Driver Machine Interface; 2005.

Comité Européen de Normalisation Electrotechnique – CENELEC. (2004). European rail traffic management system – driver machine interface. European Standard WGA9D, prEN 50XX6-1-7.

ERA Driver Machine Interface Specification version 2.3; ERTMS Unit, ERA_ERTMS_015560; 2009.

Ergonomic Data Entry Procedure for ERTMS/ETCS; Functional requirements specification; version 3.1; report 1963; September 1998.

Ergonomic study of EIRENE Sub-Project Team (EIRENE – MMI Requirements and Design Proposals (phase 5) – EPT.MMI/1236/97 Arbo Management Groep\Ergonomics 1787–5 version 2.1).

Ergonomic study of UIC (Development of a European solution for the Man–machine ETCS Interface – A200/M.F5-945222-02.00-950228).

Ergonomic study of UIC (Auditory display for ERTMS/ETCSEIRENE – MMI Phase 2 – Auditory Language; A200.1/M.6012.02-02.10-981015 Version 2.1).

Ergonomic study of UIC (Development of a European solution for the Man–machine ETCS Interface; UIC/ERRI 1996; A200/M.F5-945222-02.00-950228).

Flurscheim, C. H. (Ed.). (1983). *Industrial design in engineering, a marriage of techniques*. London: Design Council.

Kirwan, B. (1994). *A guide to practical human reliability assessment*. London: Taylor & Francis.

Metzger, U., & Vorderegger, J. (2004). Human factors und Ergonomie im European Rail Traffic Management System/European Train Control System – Einheitliche Anzeige und Bedienung. *Signal and Draht, 12*, 35–40.

Norman, D. A. (1988). *The psychology of everyday things*. New York: Basic Books.

Norman, D. A., & Draper, S. W. (Eds.). (1986). *User centered system design: New perspectives on human-computer interaction*. Hillsdale: Lawrence Erlbaum.

Pachl, J (2002) Systemtechnik des Schienenverkehrs. 3. Aufl., Teubner Verlag

Rasmussen, J. (1986). *Information processing and human-machine interaction: An approach to cognitive engineering*. Amsterdam: Elsevier.

Wickens, C. D., Gordon, S. E., & Liu, Y. (1998). *An introduction to human factors engineering*. New York: Addison Wesley Longman.

Perspectives for Future Information Systems – the Paradigmatic Case of Traffic and Transportation

Alf Zimmer

Charting a roadmap for future information systems will depend on ancillary conditions and constraints which in themselves might undergo changes.

- The future development of information systems themselves (e.g. the integration of diverse information systems by cloud computing, the integration of social and informational networks into cooperative work and live spaces sharing knowledge bases, etc.); this will necessitate novel approaches to information safety and legal safeguards for privacy
- The future developments of other organizational, societal, and technical systems, (e.g. we have to expect a higher demand for inter-modal systems in transportation for people as well as for goods, pilot-less planes, driver-less trains, or perhaps even driver-less cars)
- The future demographic development with an aging population in Europe and the Far East which has to interact with "young populations" in other regions; this entails the consequent orientation of information systems towards the demands of the user in their specific culture and situation (e.g. the integration of real and virtual mobility, that is, moving information instead of people or goods as in tele-teaching, tele-cooperative work, etc. – a good example is the development of tele-medicine systems providing general access to high quality diagnosis and therapy without moving the patient more than necessary)
- The future demands on traffic and transportation (e.g. on the one hand providing assisted mobility or virtual mobility in areas with aging populations, and on the

A. Zimmer (✉)
Lehrstuhl für Experimentelle und Angewandte Psychologie, Universität Regensburg, Regensburg, Germany
e-mail: alf.zimmer@psychologie.uni-regensburg.de

M. Stein and P. Sandl (eds.), *Information Ergonomics*,

DOI 10.1007/978-3-642-25841-1_9,

other hand in developing countries tailoring the design of traffic and information systems to the demands and competences of young populations)
- The future demands for distributed and – at the same time – integrated development and production systems (e.g. the synchronization and coordination of material and informational production chains)

In parentheses are listed foreseeable developments. However, if and when these will be realized depends on the political and legal framework for them. They are difficult to predict because they – in turn – will depend on further breakthroughs in information transmission, safety, and management.

A common theme for all these developments, however, is that they presuppose novel forms of interaction between humans and technical systems. The different aspects outlined above are not independent but constitute different perspectives on what has been coined the "System of Systems". In regard to the interaction of humans and technical systems, a system of systems has the potential to provide the user with a coherent simultaneous image of the environment as perceived by the user augmented with the information given by the technical system. A necessary precondition for this feature is the identification of the situation specific user needs for information. The cultural background of the user as well as the level of expertise plus the preferred modes of interaction, which depend on the physical state, age, expertise and culture of the users has to be taken into account to built up this coherent and augmented image of the situation.

For theses reasons, future technological developments in this field the very beginning ought to take the potential user into account for whom it is of primary importance to have his situation specific information needs fulfilled. When this is given, technology related parameters of the information processing units, like the number of floating point operations or the Baud rate are not of importance for the user. Admittedly, they are of high importance for the functioning of the system but for the user they remain in the background or even become invisible and therefore unattended. For this reason specific warning measures have to be provided for the case of system overload.

The consistent orientation towards the user needs, however, has limitations. On the one hand, the users' frames of reference in information processing have to be considered, which are the precondition for situation awareness (Endsley 1997). On the other hand, the access to user related information and its proliferation has to be regulated in order to provide the necessary protection of privacy.

Especially in the field of individual mobility this is apparent and the development of assistive technologies for cars has shown that intelligent adaptations for different populations of drivers and different tasks are not only technologically viable but also accepted by the users.

In the following, information systems for traffic and transportation are used as a paradigmatic field for a comprehensive approach. Whenever possible, it is shown how these considerations can be generalized to other fields.

1 User Oriented Information for Traffic and Transportation: A Paradigmatic Example for an Efficient Mesh of Information Technology

In order to chart the further developments of user oriented information systems for traffic and transportation it is necessary to give a systematic overview of the user's demands: **what** needs **when** and **how** to be known by the user, that is, beyond what is obvious or already known by the user. It has to kept in mind that users mostly do not organize their actions according to a rule book or a formal manual but according to perceived situational demands and earlier experiences in similar situations; this has been termed 'situated action' (Suchman 1987).

As a consequence of this, future systems have to address the following main topics: situational and individual *specificity* of information, *actuality* and *relevance* of the information, and *effectivity* of the mode of transmitting this information to the user.

2 Socio-cultural Environment and Human Competence

Future developments will depend not only on technological breakthroughs but even more on organizational, institutional, and legal developments which have to be taken into account because they influence cultural differences underlying everyday behaviour including driving. Kluckhohn (1951) has already described this as follows: "culture consists in patterned ways of thinking, feeling and reacting, acquired and transmitted mainly by symbols, constituting the distinctive achievements of human groups, including their embodiments in artifacts; the essential core of culture consists of traditional ideas and especially their attached values" (p. 86).

The cultural differences do not only play a role on the global level where technical artifacts and the concomitant information systems have to be designed in such a way that they fit into the specific cultures worldwide (for the background, see Nisbett 2003), but also on a local level, e.g. the populations in metropolitan and rural differ not only in their openness to new technologies but also in their willingness to accept the transition from individual mobility to public transportation.

If – as Kluckhohn (1951) argues – cultural differences in thinking and behaving become apparent in their relation to symbols and artefacts, the ways of transmitting information from information systems to the users will be of special importance. Actually, since about 15 years an increasing awareness for these questions can be observed in the domain of human–machine interaction or human–computer communication (Fang and Rau 2003; Honold 2000; Knapp 2009; Quaet-Faslem 2005). The cultural differences play a role in the understanding and usage of menu systems (e.g. Fang and Rau 2003; Knapp 2009) as well as in the ease of understanding commands or interpreting the meaning of icons (e.g. Honold 2000).

However, – as the results of Knapp (2009) show – the usage of new technologies, in her case route guidance systems for cars, influence the cultural development, too, that is, European and Chinese users of such systems are more similar in their behaviour than Chinese users and Chinese non-users. How the detection of cultural differences could stimulate further developments in human–machine interaction can be seen in the preference for episodic information of Chinese users in comparison with European users who usually prefer conceptual information. However, novel results in cognitive psychology show that in situations of high capacity demand on the working memory Westerners, too, use episodic information (see Baddeley 2000). It might be that the support by episodic information would have added value also for Westerners, especially for situation awareness.

3 Beyond Visual Information: Multi-modal Interactions

One important mode of human–machine interaction which is probably culturally fair is haptics (for fundamentals see Klatzky and Lederman 1992): From providing information via tactile perception ("tactons" Brown et al. 2005) to directly influencing actions through force feedback (Dennerlein and Yang 2001) there are new avenues for human–machine communication, which might be of special importance for the interface design of further information systems for traffic and transportation.

Visual and acoustical information, which is still prevalent for warning and guidance share the drawback that they are exposed to strong interference from the environment and additionally put high demands on semantic processing. In contrast, tactile information can be provided selectively, in the extreme only for the hand or the foot, which has to react upon this information (for an overview, see Zimmer 1998, 2002, or Vilimek 2007). However, even further developments in the field of tactons, which are tactile equivalents to icons, will not provide anything resembling the semantic richness of spoken or written language or pictures (for the design principles for tactons, see Oakley et al. 2002). On the other hand, many compact commands or indications for orientations as traditionally provided by icons or earcons can be presented in the tactual mode, too. Examples might be "urgency" or "acceleration/deceleration".

The major advantage of tactile information however, becomes apparent when it immediately acts upon the effectors by means of force feedback or vibrations. In these cases the reaction speed is very high, the error rates are low, and the interference with other tasks e.g. those related to driving are negligible (Bengler 2001; Bubb 2001; Vilimek 2007). That in the field of car driving the highest effect for tactile information have been found in the support for regulating behaviour, e.g. lane keeping, and somewhat less in the support for manoeuvring, e.g. overtaking. For navigating map and/or speech information will be indispensable; in some situations tactile and visual or spoken information can be integrated, e.g. during the approach to an exit.

On the one hand the focus on interface design of cars is due to the fact that the drivers are so varied and on the other hand the competition between OEMs to provide innovative forms of interfacing is extremely high. This has led to a faster development in interface design for cars than – for instance – for airplanes, trains, or ships where the operators are usually highly trained and the procedures are regulated if not constrained. However, new developments in cockpit design show that even – or especially – highly trained professionals can profit from an improved user-centred interface design (Kellerer 2010).

4 The Challenge of Interaction Design

While the user-centred design for interfaces constitutes a field of development which is already consolidating, the field of interaction design is still in an emerging phase – for this reason Norman (2010) regards it as art, not as science. This is partially due to the sheer complexity of possible interactions between information users and information sources. More important perhaps is the foreseeable development from distributed *autonomous* systems to distributed *and massively interacting* systems, prototypical examples are Google's Chrome OS which gives the user access to servers without geographical constraints or the mobile-phone based navigation systems which directly access the internet and therefore are – at least potentially – more precise and up-to-date than systems which rely on CD-ROM based digital maps. Further developments in cloud computing will not only allow access to more sources of information but will also allow "information harvesting" (comparable to "energy harvesting", where all available energy is transformed into usable energy), that is, such systems will monitor all available information which might be relevant for a specific user. For instance, in the Netherlands the Tom-Tom navigation systems utilize the fact that the motion of individual mobile phones from one cell to the next can be geographically mapped and thus allow a highly valid and timely prediction of traffic congestions.

From my point of view, the future of information systems will lie in the development of situation specific but interconnected systems, providing information that fits into the actual demands of the user. E.g. for individual surface mobility as compared to air traffic this implies a very high variability in situations due to the drivers' behaviour: as long as enforced compliance is neither technically feasible, nor legally admissible, traffic information systems have to enhance and assist the competence of the driver and not to govern or override it.

The effective mesh of operator competence and information support will depend on the quality of the information given, its specificity for the situational demands, and the fit between the mode of informing and the required actions. On the one hand, it will be the task for human-factors specialists to design interfaces and modes that allow the easy and effective integration of information relying on sensor data into the individual knowledge of the operator.

On the other hand one crucial challenge for information science will lie in the communication between the systems on the different levels, (e.g. in the case of individual traffic: the communication between systems supporting the regulation of single cars and systems managing the regional traffic flow).

Due to systems architectures and concerns about safety and privacy this kind of information integration usually does not function bottom–up as well as top down; e.g. in planning a trip one can do the entire scheduling from door to door on the computer; however the information contained in the reservations usually is not shared among the different providers for the transportation on the legs of the trip and therefore cannot be used for rescheduling in cases of delays or other kinds of perturbation. Using cloud computing as a means for efficient data harvesting could provide the support for optimized transportation in productions chains, which are resilient, that is, functioning efficiently even in the face of disruptions or perturbations.

This kind of vertical information integration in transport could also serve as a model for horizontal information integration in research and development. It is not unusual that in large corporations or consortia, responsible for the development of complex systems, not only object-related knowledge and expertise exist, but also knowledge about the interaction of such objects with its environment during its lifetime cycle. Usually this information is localized in separate departments and not easily accessible for other departments. The consequences of this lack of horizontal information sharing has become apparent when in Germany a new type of fast trains was introduced which disturbed the safety relevant electronic signalling system due to high electric field forces. However, the same corporation responsible for the production of the train had planned and installed the signalling system before.

Admittedly, for the time being there is a lack of systems providing at the same time ease of use, reliability, data safety and privacy protection. Nevertheless, the traditional goals in ergonomics, namely efficiency and ease of use, will only be attainable if the information systems are designed as integrated systems and not as collections of separate data bases, tools, and interfaces. It will be a challenge to develop architectures for these "systems of systems" which overcome the pitfalls of the traditional additive procedures in combining systems where errors in local systems can influence the functionality of the global system. With a coherent systems architecture the "system of systems" cannot only be more efficient but even more resilient in the case of external perturbations or failure of components than additive systems.

References

Baddeley, A. D. (2000). The episodic buffer: A new component of working memory? *Trends in Cognitive Sciences, 4*, 417–423.

Bengler, K. (2001). Aspekte multimodaler Bedienung und Anzeige im Automobil. In T. Jürgensohn & K. P. Timpe (Eds.), see above (pp. 195–205).

Brown, L. M., Brewster, S. A., & Purchase, H. C. (2005). A first investigation into the effectiveness of tactons. *Proceedings of World Haptics*, IEEE, Pisa, pp. 167–176.

Bubb, H. (2001). Haptik im Kraftfahrzeug. In T. Jürgensohn & K. P. Timpe (Eds.), see above (pp. 155–175).

Dennerlein, J. T., & Yang, M. C. (2001). Haptic force-feedback devices fort he office computer: Performance and musco-skeletal loading issues. *Human Factors, 43*, 278–286.

Endsley, M. R. (1997). Toward a theory of situation awareness in dynamic systems. *Human Factors, 37*, 32–64.

Fang, X., & Rau, P. L. (2003). Culture differences in design of portal sites. *Ergonomics, 46*, 242–254.

Honold, P. (2000). Culture and context: An empirical study for the development of a framework for the elicitation of cultural influence in product usage. *International Journal of Human Computer Interaction, 12*, 327–345.

Kellerer, J. (2010). Untersuchung zur Auswahl von Bedienelementen für Großflächen-Displays In Flugzeug-Cockpits. Doctoral dissertation. Darmstadt: Technische Universität Darmstadt.

Klatzky, R., & Lederman, S. J. (1992). Stages of manual exploration in haptic object recognition. *Perception & Psychophysics, 52*, 661–670.

Kluckhohn, C. (1951). The study of culture. In D. Lerner & D. Lasswell (Eds.), *The policy sciences* (pp. 86–101). Stanford: Stanford University Press.

Knapp, B. (2009). *Nutzerzentrierte Gestaltung von Informationssystemen im PKW für den internationalen Kontext*. Berlin: Mensch und Buch Verlag.

Nisbett, R. E. (2003). *The geography of thought. How Asians and Westerners think differently. And why?* New York: The Free Press.

Norman, D. A.(2010) Living with Complexity. Cambridge, Mass.: The MIT Press.

Oakley, I., Adams, A., Brewster, S. A., & Gray, P. D. (2002). Guidelines for the design of haptic widgets. *Proceedings of BCS HCI 2002*, Springer, London, pp. 195–212.

Quaet-Faslem, P. (2005). Designing for international markets. Between ignorance and clichés. *Proceedings of the 11th international conference on human–computer interaction [CD-ROM]*, Lawrence Erlbaum, Mahwah.

Suchman, L. A. (1987). *Plans and situated actions*. Cambridge: Cambridge University Press.

Vilimek, R. (2007). *Gestaltungsaspekte multimodaler Interaktion im Fahrzeug*. Düsseldorf: VDI Verlag.

Zimmer, A. (1998). *Anwendbarkeit taktiler Reize für die Informationsverarbeitung*. Regensburg: Forschungsbericht des Lehrstuhls für Psychologie II.

Zimmer, A. (2002). Berkeley's touch. Is only one sensory modality the basis of the perception of reality? In L. Albertazzi (Ed.), *Unfolding perceptual continua* (pp. 205–221). Amsterdam: John Benjamins.

Index

M. Stein and P. Sandl (eds.), *Information Ergonomics*,
DOI 10.1007/978-3-642-25841-1, © Springer-Verlag Berlin Heidelberg 2012

MIX
Papier aus verantwortungsvollen Quellen
Paper from responsible sources
FSC® C105338

If you have any concerns about our products,
you can contact us on
ProductSafety@springernature.com

In case Publisher is established outside the EU,
the EU authorized representative is:
Springer Nature Customer Service Center GmbH
Europaplatz 3, 69115 Heidelberg, Germany

Printed by Libri Plureos GmbH
in Hamburg, Germany